Estates Gazette

# Planning Law Reports 2004

Volume 1

Edited by
Barry Denyer-Green LLM PhD FRICS
of Middle Temple, barrister
and
Navjit Ubhi LLB (Hons)
of Middle Temple, barrister

Hélène Préchac
Publisher's editor

2004

Routledge
Taylor & Francis Group
LONDON AND NEW YORK

First published 2004 by Estates Gazette

Published 2014 by Routledge
2 Park Square, Milton Park, Abingdon, Oxon OX14 4RN
711 Third Avenue, New York, NY, 10017, USA

*Routledge is an imprint of the Taylor & Francis Group, an informa business*



ISBN 13: 9780728204386 (hbk)
ISBN 13: 9780728204461 (SET)

Transferred to Digital Printing in 2014

# CONTENTS



**CASES REPORTED IN THIS VOLUME**

# INDEX OF SUBJECT MATTER

## CASES JUDICIALLY CONSIDERED

**For case references see pages shown in bold type**

# Fidler *v* First Secretary of State and another

[2003] EWHC 2003 (Admin)

QUEEN'S BENCH DIVISION: ADMINISTRATIVE COURT
*Richards J*
1 October 2003

**Enforcement – "Second-bite" enforcement notice – Section 171B(4)(b) of Town and Country Planning Act 1990 – Second notice alleged material change of use in different terms from earlier notices – Whether second notice valid "second bite" notice** A

Between January 2001 and February 2002, the second respondent council served on the appellant enforcement notices, identified as notices A to F, B
alleging various breaches of planning control. In February 2002, they served a further notice, identified as notice I, alleging a material change of use (in different terms). The appellant appealed the notices, and the council's decision to refuse an application for a certificate of lawful development. The inspector made a preliminary finding that notices B, D and E, alleging various material changes of use, were invalid and could not be corrected. C
In December 2002, the inspector allowed the appeals in respect of notices B, C, D, E, F and G; the notices were duly quashed. Notices A and I were upheld. Notice I was varied and corrected so as to include a substituted description of the uses alleged to be in breach, and a substituted statement of the necessary remedial action. The inspector accepted that notice I was a valid "second-bite" notice within section 171B(4)(b) of the Town and Country Planning Act 1990, and it was therefore deemed to have been D
served within the time limit for enforcement.

The main part of the appellant's case on the appeal failed, together with the substantial dismissal of the appeal against notice I. He appealed under section 289 of the 1990 Act against the inspector's decision to uphold notices A and I, and, under section 288, sought to challenge part of the decision against him on the application for a certificate of lawful E
development. He contended, *inter alia,* that the inspector had been wrong to find that notice I was a valid "second-bite" notice, within the meaning of section 171B(4)(b), in respect of the same breaches of planning control as had been alleged in notices B, D and E, which the inspector had quashed because the alleged change of use had been very differently described. He also advanced other grounds of appeal relating to the inspector's decision F
concerning the material change of use and the application for a certificate of lawful development.

***Held*** The appeal was allowed.

Notice I was wider in substance than the earlier notices and was directed to additional facts. It did not simply state more accurately what was described in those earlier notices; it widened the scope of the notices and affected other buildings, uses and users. Section 171B(4)(b), as interpreted in *Jarmain* v *Secretary of State for the Environment, Transport*

A *and the Regions (No 1)* [2000] 2 PLR 126, did not apply to notice I; the notice was not a "second-bite" notice. The other grounds of appeal were dismissed.

**Cases referred to in the judgment**

*Belmont Riding Centre* v *First Secretary of State* [2003] EWHC 1895 (Admin)

B *Burdle* v *Secretary of State for the Environment* [1972] 1 WLR 1207; [1972] 3 All ER 240; (1972) 70 LGR 511; 24 P&CR 174; [1972] EGD 678; 223 EG 1597, DC

*Jarmain* v *Secretary of State for the Environment, Transport and the Regions (No 1)* [2000] 2 PLR 126; [2000] JPL 1063

*Jennings Motors Ltd* v *Secretary of State for the Environment* [1982] QB 541; [1982] 2 WLR 131; [1982] 1 All ER 471; (1982) 80 LGR 226; 41 P&CR 221; [1982] 1 EGLR 166; [1982] EGD 1186; 261 EG 994; [1982] JPL 181, CA

C *Kensington and Chelsea Royal London Borough Council* v *Secretary of State for the Environment* [1981] JPL 50

*Lilo Blum* v *Secretary of State for the Environment* [1987] JPL 278

*Scott* v *Secretary of State for the Environment, Transport and the Regions (Enforcement notice); Scott* v *Cheshire County Council* unreported 16 October 2000; [2000] JPL 833, QB

D **Application and appeal under Town and Country Planning Act 1990**

This was an application under section 288 and an appeal under section 289 of the Town and Country Planning Act 1990 by the appellant, Robert Fidler, against the decision of the First Secretary of State upholding various enforcement notices issued by the second respondents, Reigate and Banstead Borough Council, and their refusal to grant the claimant a E certificate of lawful development.

*Jonathan Clay (instructed by DMH, of Brighton) appeared for the appellant, Robert Fidler.*

*Timothy Mould (instructed by the Treasury Solicitor) represented the first respondent, the First Secretary of State.*

F *The second respondents, Reigate and Banstead Borough Council, did not appear and were not represented.*

*The following judgment was delivered.*

**RICHARDS J:**

[1] These proceedings, under sections 288 and 289 of the Town and G Country Planning Act 1990, concern a former farmyard and adjacent land at Honeycrock Farm, Salfords, Surrey, which is owned by the appellant, Mr Robert Fidler. The site includes a number of buildings (units 1 to 8) grouped around a southern yard and two attached steel-framed barns (units 9 and 10) in a northern yard. It is not in issue that there are now three planning units on the site: planning unit A, comprising units 1 and 2 in H the south-west corner, planning unit B, comprising essentially agricultural land at the northern end, and planning unit C, comprising the balance of the site, including units 3 to 10 and the southern and northern yards.

[2] The site has been the subject of a number of enforcement notices issued by the local planning authority, Reigate and Banstead Borough Council. Those of most immediate relevance were issued between

January 2001 and February 2002, and are referred to as notices A to G and notice I.

[3] The first of the notices, chronologically, was notice G, which was issued on 4 January 2001, and which related to the reconstruction of the roof of unit 7.

[4] The appellant then applied, on 22 January 2001, for a certificate of lawful existing use or development under section 191 of the 1990 Act in relation to various uses of the site, contending that there had been a deemed grant of planning permission by under-enforcement under section 173(11) of that Act and that, in any case, there had been 10 years' continuous use of the land for mixed commercial uses.

[5] The council did not determine that application, but, instead, issued notices A to F on 7 March 2001, alleging a variety of breaches of planning control.

[6] All the enforcement notices were appealed under section 174 of the 1990 Act, and the deemed refusal (through non-determination) of the application for a certificate of lawful development was appealed under section 195.

[7] The First Secretary of State appointed an inspector, Mr DE Morden, to determine all the appeals. A public inquiry opened on 22 January 2002, but was adjourned after four days so that the inspector could make preliminary findings as to the validity of certain of the enforcement notices and on other legal submissions made by the parties. His preliminary findings were issued on 26 February 2002, and included findings that notices B, D and E, which alleged various changes of use on the site, were invalid and could not be corrected without injustice to the appellant.

[8] Meanwhile, the council had issued a further enforcement notice, notice I, on 4 February 2002, again alleging, although in different terms, a material change of use of the site. An appeal against that notice was dealt with together with the existing appeals. The resumed inquiry in relation to all the appeals was held in May 2002.

[9] The inspector's decision letter was issued on 17 December 2002. It extended to more than 30 pages plus appendices, including details of the appeals at appendix 1 and his preliminary findings at appendix 2. The inspector allowed the appeals in relation to notices B, C, D, E, F and G, and quashed those notices. In respect of notice A, he dismissed the appeal under ground (a) of section 174(2); he stated that the appeal under ground (b) had been withdrawn, although whether that was correct is in issue. He corrected and varied notice I so as to include a substituted description of the uses alleged to be in breach, a substituted statement of the remedial action required, and a period of 12 months, rather than six months, for compliance. He then dismissed the appeal against notice I, except to the extent of permitting certain uses of units 6, 7 and 8. He allowed, to a limited extent, the appeal in respect of the certificate of lawful development, which he described as appeal H, and granted a certificate relating to certain limited uses; but the main part of the appellant's case on the appeal failed, together with the substantial dismissal of the appeal against notice I.

A [10] The appellant now appeals to this court, under section 289 of the 1990 Act, against the inspector's decision to uphold notices A and I. He also applies, under section 288, to challenge that part of the decision in which the inspector had found against the appellant on the application for a certificate of lawful development.

[11] The main issues raised before this court are whether: B (i) the inspector had been entitled to find that notice I was a valid "second- bite" notice within section 171B(4)(b) of the 1990 Act (the second-bite issue); (ii) he had erred in finding that there had been a material change of use of the land (the material change of use issue); (iii) he had erred in relation to the question of whether various commercial uses of the site had been carried on long enough to become lawful (the previously C established lawful use issue); (iv) he had erred in various respects in his dismissal of the appeal against notice A (the notice A appeal issue); and (v) he had erred in his rejection of the appellant's case that the combined effect of section 173(11) of the 1990 Act and an enforcement notice issued by the council in 1996 had been to confer deemed planning permission on uses that the notice did not require to cease (the deemed planning permission D issue). Various other matters raised in the appellant's written or oral submissions can be dealt with, so far as necessary, when considering those main issues.

[12] The argument before me has been between the appellant and the First Secretary of State. The council were not represented at the hearing.

*Statutory framework*

E [13] Part III of the 1990 Act is concerned with control over development. By section 55(1), "development" means the carrying out of building, engineering, mining or other operations in, on, over or under land, or the making of any material change in the use of any buildings or other land. By section 55(2), however, certain operations or uses of land are not to be taken to involve development. That includes, under section F 55(2)(a), certain works of maintenance, improvement or alteration that (i) affect only the interior of the building or (ii) do not materially affect the external appearance of the building. It also includes, under section 55(2)(f), in the case of buildings or other land that are used for a purpose of any class specified in an order made by the Secretary of State, the use of the buildings or other land for any other purpose of the same class. G The classes relevant to this case are Classes B1 (business), B2 (general industrial) and B8 (storage or distribution), as specified in the Town and Country Planning (Use Classes) Order 1987.

[14] By section 57, subject to the detailed provisions of that section, planning permission is required for the carrying out of any development of land.

H [15] Part VII deals with enforcement. Section 171A contains basic provisions:

> .–(1) For the purposes of this Act –
>
> (a) carrying out development without the required planning permission;...

> ... A
>
> constitutes a breach of planning control.
>
> (2) For the purposes of this Act –
>
> (a) the issue of an enforcement notice (defined in section 172)...
>
> ...
>
> constitutes taking enforcement action.

[16] Time limits for enforcement action are contained in section 171B. B Subsection (1) lays down a four-year limit for breaches consisting of the carrying out without planning permission of building, engineering, mining or other operations. Subsection (2) lays down a like period for a breach consisting of change of use of a building to a single dwellinghouse. The focus for present purposes, however, is on subsections (3) and (4)(b), which lay down a 10-year period for all other cases, and allow an C additional four-year period for further enforcement action in certain cases (the second-bite provision). They read, so far as material:

> (3) In the case of any other breach of planning control, no enforcement action may be taken after the end of the period of ten years beginning with the date of the breach.
>
> (4) The preceding subsections do not prevent – D
>
> ...
>
> (b) taking further enforcement action in respect of any breach of planning control if, during the period of four years ending with that action being taken, the local planning authority have taken or purported to take enforcement action in respect of that breach.

[17] General provisions governing the issue of an enforcement E notice and the contents of a notice are contained in sections 172 and 173. Section 173(11), upon which the appellant relies in support of the application for a certificate of lawful development, provides:

> (11) Where –
>
> (a) an enforcement notice in respect of any breach of planning control could have required any buildings or works to be removed or any activity to F cease, but does not do so; and
>
> (b) all the requirements of the notice have been complied with,
>
> then, so far as the notice did not so require, planning permission shall be treated as having been granted by virtue of section 73A in respect of development consisting of the construction of the buildings or works or, as the case may be, the carrying out of the activities.

G

[18] The right of appeal to the Secretary of State against an enforcement notice is governed by section 174. By section 174(2), the grounds upon which an appeal may be brought include:

> (a) that, in respect of any breach of planning control which may be constituted by the matters stated in the notice, planning permission ought H to be granted...;
>
> (b) that those matters have not occurred;
>
> ...
>
> (d) that, at the date when the notice was issued, no enforcement action could be taken in respect of any breach of planning control which may be constituted by those matters.

[19] On an appeal under section 174, the Secretary of State may, by section 176(1), correct any defect, error or misdescription in the enforcement notice, or vary the terms of the enforcement notice, "if he is satisfied that the correction or variation will not cause injustice to the appellant or the local planning authority".

[20] Applications for a certificate of lawful development are governed by section 191. The central provisions are subsections (4) to (6):

> (4) If, on an application under this section, the local planning authority are provided with information satisfying them of the lawfulness at the time of the application of the use, operations or other matter described in the application, or that description as modified by the local planning authority or a description substituted by them, they shall issue a certificate to that effect; and in any other case they shall refuse the application.
>
> (5) A certificate under this section shall –
>
> (a) specify the land to which it relates;
>
> (b) describe the use, operations or other matter in question (in the case of any use falling within one of the classes specified in an order under section 55(2)(f), identifying it by reference to that class);
>
> (c) give the reasons for determining the use, operations or other matters to be lawful; and
>
> (d) specify the date of the application for the certificate.
>
> (6) The lawfulness of any use, operations or other matter for which a certificate is in force under this section shall be conclusively presumed.

[21] Section 195 confers a right of appeal against a local planning authority's refusal of a certificate or their failure to determine the application within the time laid down.

*(1) Second-bite issue*

[22] The issue here is whether the inspector had been correct to decide that notice I was a valid second-bite notice, involving further enforcement action within the meaning of section 171B(4)(b) in respect of the same breaches of planning control as had been alleged in notices B, D and E, which the inspector had quashed.

[23] The effect of treating notice I as a second-bite notice was to push back the time limit for enforcement action to 10 years before the date of those earlier notices, that is, to 7 March 1991, rather than limiting it to 10 years before the date of notice I itself, namely to 4 February 1992. This was potentially important, since the inspector had found changes during the period from March 1991 to February 1992, and had taken them into account when determining whether the mixed use on site as at the date of notice I had come about through a material change of use.

[24] I have already set out the terms of section 171B(4)(b), which relate to the taking of enforcement action "in respect of any breach of planning control" if, during the relevant period, the authority have previously taken, or purported to have taken, enforcement action "in respect of that breach".

[25] The leading case on that provision is *Jarmain* v *Secretary of State for the Environment, Transport and the Regions (No 1)* [2000] 2 PLR 126. The brief facts were that planning permission had been granted to the

appellant for use of a mobile home on land, subject to a condition whereby the permission was limited in time. Before the permission expired, he had changed his accommodation from a mobile home to a permanent single-storey dwelling, the permanent dwelling having been constructed around a mobile home that had been damaged while being installed on the site. Following expiry of the temporary permission, in March 1996, the local planning authority issued an enforcement notice, alleging breachof condition in that the mobile home remained on the land. In March1998, that enforcement notice was withdrawn and a second notice was issued, in which the breach of planning control alleged was erection of a single-storey building without permission. The relevant issue was whether the second enforcement notice fell within section 171B(4)(b).

[26] The inspector had found that both enforcement notices were directed at the same structure and that, although the actual description of the breach differed, in issuing the second notice the authority had purported to take enforcement action in respect of the same breach as they had purported to do under the first notice. The deputy judge upheld the inspector's conclusion, saying, at pp133G-134F:

> When section 171B is read as a whole it appears clear that the section is concerning itself with breaches of planning control and the time-limits applicable when enforcing against these breaches.
>
> Various types of planning control relating to different types of development are referred to in the section. What the section is concerned with are the various acts of development that constitute the breaches of control, not the description of the substantive activities. It is the subject-matter of the enforcement notice, the actual development that is being referred to by the words "that breach", rather than the words that have been used to characterise the development, that constitutes the breach of planning control. The description of a breach cannot in itself be a breach of planning control. What this proper interpretation means is that the subsection cannot be used to cover two different physical developments or two different changes of use, but it can be used to cover the same actual breach of development control, which is described in different ways.
>
> This interpretation is clear enough from the wording of the section, but is also apt for illustration by the facts of this case. The facts of the situation are that a permanent dwellinghouse had been constructed without planning permission by 12 April 1993. That constituted a breach of planning control. Such a breach is identified in section 171B(1). The local planning authority purported (note that that is the word used in section 171B(4)(b)) to enforce against that physical development, a dwellinghouse construction, but, mistakenly, did not appreciate that the development carried out was a permanent dwellinghouse; they therefore described the breach as a mobile home still on site in breach of the time-limit condition. That mistake did not change the nature of the breach of planning control, which was still construction of a permanent dwellinghouse without planning permission, and it was plainly against that physical development that the local planning authority had purported to enforce. When they learned of their mistake they correctly described the breach of planning control in a new enforcement notice, but it was still the same physical development construction that had always been the only breach of planning control under consideration, and,

> A therefore, the second enforcement notice was enabled by the subsection and the inspector was correct in taking this approach.
>
> This interpretation is also consistent with the purposes of the section which, as can be seen in the general note to the subsection in Vol 2 of the *Planning Encyclopaedia* (2-3598/1), removed the protection given to developers in previous legislation who succeeded in establishing technical errors in enforcement notices while time continued to run, so that cases arose when B time for service of an enforcement notice without the defect had elapsed.

[27] Before the Court of Appeal, the Secretary of State contended that the deputy judge's reasoning was correct. Counsel for the appellant submitted that the breach that founds the first enforcement notice must be the same as the breach that founds the second enforcement notice, and C that, in the instant case, there were entirely different breaches alleged: in the first notice, a change of use of the land in breach of the condition in the temporary permission; in the second, the carrying out of a building operation without permission. The Court of Appeal favoured the reading of the section adopted by the inspector and the deputy judge.

[28] In giving his reasons, Roch LJ placed weight upon avoiding a D situation in which notices might be defeated on technical grounds. He stated that the reading of the section for which counsel for the appellant contended "would lead to a return to the technicalities that bedevilled the enforcement of planning control prior to the passing of the 1990 and 1991 Acts": see p136H.

[29] So, too, Brooke LJ contrasted the appellant's purist approach with E the pragmatic approach adopted by the inspector and deputy judge and, in preferring the latter, laid emphasis upon the desirability of avoiding technicalities. He stated, at pp139C-140B and 141B-D:

> Anyone who had any experience of the operation of the former law relating to the enforcement of planning control knows that it was disfigured by time-consuming litigation over technicalities, raised F by determined litigants who sought to evade the effects of enforcement action taken against them by local planning authorities on behalf of their local communities.
>
> From time to time, there were judicial explosions on the topic. Among these, the most colourful was that of Templeman J in 1974, and the most resigned that of Mr Graham Eyre QC, sitting as a deputy high court judge in 1988.
>
> G In *Eldon Garages Ltd* v *Kingston upon Hull County Borough Council* [1974] 1 All ER 358, Templeman J ended his 18-page judgment at p375B with these words:
>
> "The result of this submission, if I acceded to it, would be that the enforcement notice is a kind of spell by a witch doctor and unless the witch doctor gets the exact words of the incantation right, then the spell does not work. Well, H although, as the House of Lords pointed out, one has got to be very careful in these cases because of the individual rights involved, we have not yet got to that stage and I do not propose to begin getting near that stage... That being so, it seems to me that I must dismiss this summons. This is not an enforcement notice which relates to Bleak House, and although counsel for the plaintiffs has argued this case very skilfully and has put forward every argument open to him, I intend no disrespect to him and no criticism of

anybody when I say that it is time that the pettifogging was stopped and I dismiss this summons." A

With a wearied resignation born of a professional lifetime of experience of all the difficulties posed by the existing law, Mr Graham Eyre QC said in *West Oxfordshire District Council* v *Secretary of State for the Environment* [1988] JPL 324 at pp324-5:

"It... [is] perhaps remarkable that nearly four decades [have]... passed since the enforcement notice machinery first emerged in the Town and Country Planning Act 1947 during which time the legislature [has] made substantial amendments to the statutory provisions so as to remove or substantially reduce the powers of the courts to interfere on technical grounds, and yet this court [has]... been treated to a rehearsal of somewhat arid technicalities most of which [have]... a ring of nostalgia, in its true sense, and largely unwelcome familiarity." B

I am quite satisfied that one of parliament's main purposes in 1991, in overhauling Part II of the 1990 Act, was to spare those like Mr Eyre the pain of returning to those arid technicalities... C

...

It appears to me that if we were to allow this appeal, we would be in danger of allowing enforcement law to return, in part, to the world of pettifoggery and arid technicalities that attracted such strong judicial disapprobation in the 1970s and 1980s. In my judgment, this court ought to be very slow to depart from the approach to the interpretation of section 171B(4)(b) that was adopted by a deputy judge with great experience in the planning field. He was satisfied that the relevant breach of planning control consisted of the erection of the unauthorised structure in the position marked on the plan attached to the enforcement notice. The council purported to take enforcement action in respect of that breach by their first notice, but failed to do so because they misdescribed the breach. They were therefore entitled by section 171B(4)(b) to take further enforcement action because four years had not elapsed since the date of the earlier purported action. D E

[30] The concerns underlying the Court of Appeal's conclusion are, however, perhaps best illustrated by Sedley LJ's judgment, at pp141E-142A: F

For the reasons given by Roch and Brooke LJJ, I agree that this appeal fails. To allow it would be to set a premium once more on technical challenges to enforcement notices that have misled nobody. It would also, in a case like the present one, be to reward conduct that, at best, was devious and, at worst, deceitful. As is apparent from the account given by Roch LJ, the appellant continued to claim that he had a mobile home on the site for as long as it suited him to do so. Then, when he thought the time was up for enforcement, he asserted that it was a permanent structure. When one looks in the unaccountable absence of photographs at his account of what he had done to the mobile home, it becomes easier to see why the council's officials were taken in. He had, by his own account, set the structure on timbers, thereby presumably elevating it from the ground; he had put on a board and felt roof; and he had retrieved other, unspecified, materials "for use in the construction". G H

It would make a farce of the legislation if, by claiming that one unlawful use was a different unlawful use, a landowner could take advantage of the four-year limit and acquire immunity from process for something he had known all along was unlawful. Miss Lieven's submission has satisfied me

A
> that to read the legislation as eschewing form in favour of substance does not give a slack local authority, at least in certain instances, an unintended length of time in which to get their notice right. The second notice must not only relate to the same facts as constitute the breach to which the first notice related, but must be served within the period for which the first one is to be taken to have been good, namely four years.

B [31] In order to assess the rival submissions as to the application of the approach in *Jarmain* to the circumstances of the present case, it is necessary, first, to compare the contents of the earlier notices (notices B, D and E) with the contents of the later notice (notice I), and then to examine the inspector's reasons for quashing the earlier notices and for treating the later notice as a valid second-bite notice.

C [32] Notice B was expressed to relate to land comprising effectively the the whole site, but the alleged breach was: "Without planning permission, change of use of the land involving the use of the building hatched black on the attached plan for industrial purposes and commercial storage purposes." The "building hatched black" was units 9 and 10.

[33] Notice D was also expressed to relate to land comprising effectively
D the whole site. The alleged breach was: "Without planning permission, change of use of the land involving the use of Unit 8 (shown hatched black on the plan) for the storage, crushing and recycling of aluminium."

[34] Notice E was likewise expressed to relate to land comprising effectively the whole site. In this case, the alleged breach was: "Without planning permission, change of use of the land to use for the open storage
E of materials not ancillary to the agricultural use of the land."

[35] Notice I was expressed to relate to land comprising only part of the site, namely planning unit C: that is, an area including units 3 to 10 and the southern and northern yards. The alleged breach was: "Without planning permission, change of use of the land to a mixed use for Class B1, B2 and B8 uses, agriculture, hay and straw dealing, haulage depot and building
F and demolition contractor's depot."

[36] The inspector, however, corrected notice I so that the alleged breach was change of use to a mixed use "consisting of agriculture, agricultural contracting, hay and straw dealing, storage (including the storage of helicopters), a vehicle haulage operator's depot, a building and demolition contractor's depot, motor vehicle servicing and repairs (including jet skis),
G courier businesses involving storage and distribution and a carpentry business": see para 168 of the decision letter. He considered the issues before him on the basis of the notice so corrected.

[37] The inspector's reasons for finding that notices B, D and E were defective and could not be corrected without injustice are set out in paras 27 to 29 of his decision:

H
> 27. The problem, in my opinion, is in the allegations as set out in the Notices; each alleges a "change of use of the land involving the change of use of..." either a particular building or all the open area outside of the buildings. I consider the Notices are not invalid on their face but the allegations are too vague to be upheld as issued. They do not specify what material change of use has taken place nor include all the uses that are

> taking place within the planning unit; be it the one as delineated on the plan A attached to the Notices B, D and E or the smaller area delineated on Notice I (which both parties agreed, at the resumption of the inquiry in May, was the correct planning unit at which all the Notices should be aimed).
>
> 28. As to the correction of the Notices, clearly I have wide powers of correction and can do so as long as there would be no injustice. Submissions were made regarding this and the possibility of amalgamating the Notices at the inquiry, in writing during the adjournment and also in closing B submissions. The appellant argued that if I corrected the Notices there would be people/businesses who had not appealed against them because they did not think the Notices concerned them and they could not appeal now. Additionally, there were buildings (eg units 6 and 7) and some users of the open yard not covered by the Notices as issued but they would be affected if the Notices were corrected to include all the relevant uses in the description C and, consequently, the requirements.
>
> 29. The Council in closing submissions accepted this point but submitted that the issue of Notice I had resolved the matter as everyone had now been served with a copy of the Notice and had the opportunity to lodge an appeal. Further, it was submitted that as the intention of Notice I was the same as B, D and E there would be no injustice if they were corrected, nor if they were amalgamated into one Notice. Whilst I agree that the site owner/appellant D could not really be in any doubt as to what the Notices were aimed at, it does not get over the fundamental point that the Notices are vague. In correcting them either individually or by amalgamation, I consider that their scope would be widened; other buildings, uses and users would be affected by them and I conclude, therefore, that in those circumstances there would be injustice. I will, therefore, quash Notices B, D and E.

E

[38] His summary of the arguments relating to the status of notice I, and his reasons for holding that it was a valid second-bite notice, are set out in paras 30 to 32:

> 30. The appellant argued that Notice I could not be considered as a second bite notice as it was for a fundamentally different allegation. I acknowledge that the area included in Notice I is different and also that the allegation itself F following the general agreement reached at the inquiry about how it ought to be corrected (and set out in Section *Ac* above) is different. There is also the possibility that it needs further correction after considering the evidence in detail on the relevant ground (c) and (d) appeals.
>
> 31. On the first two days of the inquiry there were numerous legal submissions including those regarding the validity of the Notices B, D and E in particular. I am clear from all the evidence put forward that the Council G were trying to achieve the cessation of all those uses at the site which did not have planning permission or could not be proved to be lawful and I have no doubt that the appellant (who was also the site owner) was aware that the notices were issued with that intent. I have concluded that the Council had not made the right allegations and I could not correct them without causing some injustice. However, both during the opening two days of the inquiry, H when the Council, during an adjournment, tried to produce an amalgamated Notice and since then the Council have tried to get the allegation right. That resulted in the issue of Notice I and whilst as originally worded it was still unacceptable, I consider it can be corrected (as set out in Section *Ac*).
>
> 32. In these circumstances I consider that it does purport to relate to the same breach of planning control even though the uses in Units 6 and 7 were

> A not included in the earlier Notices; some of the occupiers were also not included in the earlier Notices but I am satisfied that the uses were, bearing in mind the wide reaching wording of the allegations. I will therefore, deal with Notice I on the basis that it is a "second bite notice" and, in my opinion, it supersedes Notices B, D and E (which, I have already concluded should be quashed) wording correctly the allegations that those three Notices were aimed at.

B

[39] For the appellant, Mr Jonathan Clay submitted that there was an inconsistency between the inspector's finding that the earlier notices could not be corrected, individually or by amalgamation, without injustice because it would widen their scope, and his finding that the later notice related to the same breach of planning control as the earlier notices. In
C any case, it had not been properly open to him to find that the later notice did relate to the same breach of planning control, or to the same facts, as the earlier notices. The inspector had also relied, erroneously, upon the council's intentions, that is, what the council were trying to achieve by the earlier notices, rather than upon the breaches actually alleged in them. Thus, in applying section 171B(4)(b), he had either misdirected himself or
D reached a conclusion not reasonably open to him. His correction of notice I did nothing to remedy the problem.

[40] For the First Secretary of State, Mr Timothy Mould submitted that it was plain from the decision letter that the inspector's approach had been consciously that laid down by the Court of Appeal in *Jarmain*. Thus, there had been no misdirection. The conclusion he reached had been reasonably
E open to him on the facts, as can be seen from the relevant paragraphs of the decision, read against the background of his preliminary findings (in particular at sections 4 and 5 of appendix 2). The breach alleged in notice I was change of use of planning unit C to a mixed use of the whole site. As to the concept of a mixed use, see *Burdle* v *Secretary of State for the Environment* [1972] 3 All ER 240, at 244d, where Bridge J described the case
F of a composite (that is, mixed) use of a planning unit "where the component activities fluctuate in their intensity from time to time, but the different activities are not confined within separate and physically distinct areas of land". The inspector's approach in correcting the notice, so as to specify all the components of that mixed use, had been correct and in accordance with the guidance given in para 2.10 of Circular 10/97 *Enforcing planning*
G *control*. He had considered the earlier notices to be flawed because they took a piecemeal approach, targeting individual components of the mixed use, rather than the mixed use itself, and he had found that they could not be corrected without injustice because they were vague, they did not include all the components and other occupiers had not been served. But the fact was that, by those earlier notices, the council had been targeting
H the same breach of planning control, namely the mixed use, although it was described in different ways. The notices had been seeking to achieve the cessation of the mixed use of the land.

[41] Mr Mould further submitted that the failure in the earlier notices to embrace the mixed use as a composite whole was a technical flaw of a kind that the second-bite provisions are apt to deal with. As a matter of principle,

there can be a valid second-bite notice even where earlier notices are A incapable of being corrected without injustice. If the second-bite provisions applied only where earlier notices could be corrected without injustice, the provisions would not be needed. The appellant's case, he suggested, comes down to asking the court to rule that the *Jarmain* approach is limited to cases of misdescription of precisely the same underlying facts and does not embrace a second-bite notice which incorporates within the alleged B breach more extensive activities and a wider area than covered by earlier notices. But *Jarmain* should not be given such a narrow application. The policy considerations that caused the court to adopt the approach it did in *Jarmain* apply with equal force here. It is particularly likely in a complex mixed-use case that the local planning authority will fail to cover the full extent of a mixed use or of the area within which it is carried out, and it C would be surprising and unsatisfactory if the authority were prevented from relying upon the second-bite provisions to get it right.

[42] The issue addressed by those rival submissions is the one issue of potential importance in the case, and, fortunately, it is less bedevilled by factual detail than the rest of the case. For the reasons given below, and with a considerable degree of reluctance, I have concluded that the D appellant's submissions on it are well founded.

[43] In broad terms, it is no doubt true, as the inspector had found, that the council were trying, by the earlier notices, to achieve the cessation of all those uses at the site that did not have planning permission or could not be proved to be otherwise lawful, and that the appellant was aware that the notices had been issued with that intent. But, in considering the E application of section 171B(4)(b), I think it necessary to focus, not upon the council's broad intention but upon the terms of the enforcement notices themselves, and, in particular, upon the actual breach or breaches of planning control alleged in them.

[44] The earlier notices were directed in terms at a number of specific activities or uses alleged to give rise to a material change of use. In the F case of notice B, the alleged breach was a change of use involving use of units 9 and 10 for industrial and commercial storage purposes. In notice D, it was a change of use involving use of unit 8 for the storage, crushing and recycling of aluminium. In notice E, it was a change of use of the entire site to use for the open storage of materials not ancillary to agricultural use.

[45] It may be that, as the council suggested in their submissions to G the inspector, the notices were all targeting component parts of a mixed use, although that is not how it is put in the notices themselves. The problem, however, is that, even if viewed collectively, the notices cannot be said to have targeted the whole of the mixed use at which notice I was directed. The breach alleged in notice I is a change of use of planning unit C to a mixed use with multiple components, as described H in the original notice and as corrected by the inspector. That encompasses a wider range of components than the aggregate of uses covered by the earlier notices.

[46] The disparity between the earlier notices and notice I is inherent in the inspector's own finding, so that to correct the earlier notices so as to

cover all the uses said to be taking place on the planning unit would be to widen the scope of the notices and to affect other buildings, uses and users. It may be that most of the components of the mixed use alleged in notice I were covered by the earlier notices, and, possibly, that the most important components were so covered: the inspector mentions, in particular, the uses in units 6 and 7 as not being included in the earlier notices. But there still exists a material disparity.

[47] It follows, in my judgment, that notice I goes wider in substance than the earlier notices and is directed at additional facts. It does not simply describe more accurately what was misdescribed in the earlier notices; it goes further than the earlier notices. That takes the case outside the scope of the reasoning in *Jarmain*. Although I am prepared to accept that the *Jarmain* reasoning is capable of applying, in principle, where a single second-bite notice covers breaches alleged in a number of earlier notices, it does not, in my view, apply where the breach alleged in the new notice goes wider in substance than the totality of breaches alleged in the earlier notices.

[48] I acknowledge the force of the policy considerations referred to in *Jarmain*, upon which Mr Mould put particular weight in his submissions. It seems to me, however, that to hold that section 171B(4)(b), as interpreted in *Jarmain*, does not apply to notice I is not to allow technicality and form to prevail over substance. The problem goes beyond one of technical defects in the earlier notices. Even if the council had intended, by those earlier notices, to target the entirety of the mixed use on the site, the notices themselves fell materially short of doing so, whether viewed individually or collectively. That falls outside the range of circumstances at which section 171B(4)(b) is aimed.

[49] Accordingly, I conclude that the inspector had erred in holding notice I to be a valid second-bite notice. It does not matter, for present purposes, whether the error is described as a misdirection of law or the making of a finding not reasonably open on the facts.

[50] I have indicated that it is a conclusion that I reach with a considerable degree of reluctance. That is because, subject to any further submissions from counsel about the exercise of the court's discretion, the consequence of this error would seem to be that the decision must be quashed. That is particularly unfortunate in a case where it seems likely that the inspector's basic conclusion would have been the same if he had confined his attention to the 10 years preceding the date of notice I, and where the decision was otherwise a thorough and impressive piece of work in relation to which, as explained below, I reject all the other grounds of challenge.

[51] In case I am wrong in my conclusion on the second-bite issue, and since the rest of the case was the subject of full argument before me, I propose still to deal with the other issues.

*(2) Material change of use issue*

[52] In considering the appeal against notice I under ground (d) of section 174, the inspector found that the existing mixed use on the site had come about through a material change of use from the use that had been

there on 7 March 1991. The issue is whether he had erred in his approach to material change of use.

[53] The relevant section of the decision letter is lengthy: see paras 58 to 109. The inspector proceeded by summarising the parties' basic contentions (paras 61 to 62), considering the evidence of actual use and activity on the site up to the start of 1991 (paras 63 to 71), and since the start of 1991 (paras 72 to 85), assessing, in the light of his findings, the changes since the start of 1991 (paras 86 to 93), and, finally, expressing his reasoned conclusions on the materiality of the changes during the relevant period: paras 94 to 109.

[54] As regards the situation at the start of 1991, the inspector concluded, as a matter of fact and degree, that there were two distinct planning units within the one unit of occupation, with the southern yard in agricultural use and the northern yard in mixed use: para 70.

[55] He identified a number of changes since the beginning of 1991. Among other matters referred to was that the southern yard had also come into mixed use and the two planning units had become merged into a single planning unit with a mixed use: para 75. In mid-1991, the appellant erected unit 9 as an extension to unit 10, and also extended the area covered by the planning unit, taking in some land at the northern end of the site: para 76. In 1995, an area of land immediately to the north of unit 9, which had previously been in agricultural use, was taken into the planning unit and was used, thereafter, for commercial purposes: para 77. There were physical alterations to buildings, including, in particular, to units 9 and 10: paras 78 to 82. There were also various changes to the actual uses taking place on the site and to the occupiers: paras 83 to 85.

[56] In his assessment of the changes since the start of 1991, the inspector stated, first, that, in his view, the correct approach should be to take the planning unit as it existed in March 1991 and then look at all the physical changes since then, as well as the changes in uses and occupiers. The following are extracts from paras 87 to 93, dealing with those changes:

> From March 1991 the area used for the various uses increased by just over 50% by March 2001, in my view a considerable increase. In terms of the appearance of the land it also resulted in open storage taking place on land to the north of the building… [T]he size of the main covered area was basically doubled when unit 9 was added to unit 10 later in the same year and… it was generally used for non agricultural purposes from day one… Then in 1998 the whole structure was virtually rebuilt… and it again changed totally in character in my view, particularly in how it could be used. What had been basically a "Dutch barn" with open ends and some corrugated sheeting on the two side elevations of a plain steel framework was turned into a purpose built industrial type building (despite the corrugated cladding on the outside which can also be found on a modern agricultural building) with block work walls and partitions, including some first floor sections, and much of it with a proper concrete floor. … The changes in the type of space offered for rent after the 1998 alterations, in my view, facilitated a change in the type of use there. It changed from a basically "open" covered area under the roof of units 9 and 10… to one that provided secure, weatherproof storage on a large scale that would not have been possible in the old units 9 and 10. Further,

> it has resulted in the occupation by 2 companies specialising in the delivery of parcels which have more vehicles and staff on site than most and the establishment of a company… which also has many staff and vehicles on site… It was not disputed that these three occupiers had a higher generation of traffic to and from the site than others nor was the answer… that there had been more than a doubling of traffic movements at the site in the three years 1999 to 2002… This has also resulted, in my view, in a materially different character of use at the site… I acknowledge that the number of occupiers has remained fairly constant over the relevant period… I also consider that… the various uses undertaken by those occupiers have not, in principle, been different in terms of the number of uses that have been taking place at any one time and the general description of what they did… Some uses that were there at the start of 1991 ceased; others commenced during the 1990s and some were there for just short periods… The sorts of vehicles, materials and equipment stored/parked at the site by them was not that different to other existing occupants… so there was not, in my view, a material difference there. However, I consider that the situation changed after 1998 when the major building alterations were carried out…

[57] There followed the inspector's conclusions. I think it necessary to quote paras 94 to 100 in full:

> 94. In my view, there were fundamental changes at the appeal site between March 1991 and March 2001. These were the increase in the land area covered by the use; the increase in building mass and, therefore, the consequent increase in the amount of usable covered floor space; the large increase in the area of outside storage; the virtual rebuilding of units 9 and 10 almost as purpose built commercial buildings; the nature of the more recent occupiers of the site, in particular, of the buildings, and also the level of traffic associated with the various uses and occupiers of the site particularly the more recent tenants.
>
> 95. The changes summarised above have resulted in a mixed use that is undertaken on a much larger area. The mixed use in March 2001 was not actually on the same site at all as that in March 1991 but on a land area that had grown by over 50%, the increase in area taking place through increases on two different occasions. The new area taken into the planning unit had been in agricultural use and, in my opinion, the planning unit as it was in March 2001 had only existed since late 1994 or early 1995 at the earliest. Even if my conclusion is wrong on that as a point of law, ie, the simple increase in the size of the unit is not sufficient to preclude a success on ground (d), there are still the other changes to take into account.
>
> 96. The first increase in the size of the planning unit in 1991 was also large at about 0.1 hectares but that increase in area facilitated the building of the large extension that became known as unit 9; about 40% of its floor space was in the extended site. Taken together with the unit 9 extension, my opinion is that the site, even as first enlarged, late in 1991, resulted in a material change in the appearance of the land through the increased level of outside storage, the increased use taking place in an enlarged building and the alterations to the building itself which had approximately doubled in size with an extension of nearly 1100 square metres floor space.
>
> 97. There were then the further changes to the buildings in 1998. Units 9 and 10 were virtually rebuilt around their framework skeleton and some of the existing sheeting/panelling that formed the walls and roof. They became what I would describe as a complete building with four solid walls

and a roof, numerous solid internal partition walls and even some first floor accommodation. Use of the site, in my view, became far more intensive because of these changes as I believe the photographs taken over a long period showed, particularly regarding the level of storage both outside and inside the buildings.

98. The 1998 changes to the buildings and the different occupiers that came, in my view, as a result of the more permanent and secure space available, also led (and still does) to a far higher level of traffic to and from the site than experienced before, both in terms of staff and business vehicles. This mainly came through the introduction, not long before the issue of the Notice I, of some different types of uses (in particular the corporate games organiser and the two courier companies) with materially greater traffic generation than the uses that had been there before.

99. Whilst I accept that it might not be possible to say that any one of these changes by itself resulted in the material change in the use of the site, I consider, as a matter of fact and degree, that the use at 7 March 2001 was materially different to that which had existed on 7 March 1991. It was not just a more intensive use of the site but one where the activities making up the uses, the land area used; the extent of storage, both inside and outside; the buildings in which the uses were being undertaken and the level of traffic generated, if all taken together, describe a use that is materially different in character and nature.

100. I conclude, on the basis of the evidence that was put before me, that those changes were so significant that the mixed use there at the time the Notice was issued had, by its very different nature and character, come about through a material change of use from the use which had been there on 7 March 1991.

[58] He went on, in paras 101 to 103, to deal with points relevant to the precise terms of the correction to notice I. In paras 104 to 105, he considered whether there were any particular uses that could be regarded as lawful and should be excluded from the scope of the requirement to cease activities on the site. He found that, apart from agriculture, the only uses that should not be required to cease were hay and straw dealing and agricultural contracting. In para 106, he set out his conclusion that the appeal against notice I on ground (d) of section 174 had failed.

[59] It is convenient to mention, at the same time, how the inspector had dealt with the section 195 appeal in relation to the application for a certificate of lawful development. He pointed out, in para 107, that the appeal concerned a slightly different time period, going back 10 years from 22 January 2001 (the date of the application). He acknowledged, in the same paragraph, that he had the power, by virtue of section 191(4), to issue a certificate in terms different from those applied for where a smaller area of land for the use applied for could be identified, or where some similar but different use was found to be lawful on all or part of the site. In para 108, he found no material difference arising from the fact that the relevant date of 22 January 1991 was earlier than applicable to notice I. In the light of his conclusions on the ground (d) appeal against notice I, he found, in para 109, that the only lawful uses were agricultural contracting and hay and straw dealing, and that a certificate should be granted for those two uses.

[60] For the appellant, Mr Clay submitted that the inspector had adopted the wrong approach to the question of material change of use and had erred in finding a change of use by virtue of a change in the extent of the planning unit, the construction of new buildings and intensification of use.

[61] As to change in the extent of the planning unit, Mr Clay submitted that the inspector had erred in finding, in the first place, that there had been two distinct planning units at the start of the period. There was no physical separation between the northern and southern yards and there was no difference in circumstances that could justify a finding that the two yards constituted separate planning units in 1991, yet formed a single planning unit, planning unit C, in 2001.

[62] But even if that finding had been open to him, it is submitted that the inspector's focus should have been on the activities in the northern yard, and he had erred in taking into account the extension of the planning unit by the inclusion of the southern yard and of land to the north of unit 9. An extension of the planning unit is not in itself a breach of planning control. Although he had been entitled to find that the areas into which the commercial uses had extended could not benefit from lawful use rights, that did not help him to answer the question of whether there had been a material change of use of the area where commercial activities had been carried on since the beginning of the relevant period. If there were no material change of use of the northern yard, then the appellant would be immune from enforcement action in relation to that area. The inspector seems to have appreciated this in his comments on the application for a certificate of lawful development (para 107), but had failed to approach the matter in this way in his analysis of material change of use.

[63] As to the construction of new buildings, Mr Clay submitted, first, that the inspector's finding that a new building had been constructed was based upon a mistaken inference. In para 79, the inspector stated that the appellant withdrew his ground (b) appeal against notices A and G and "has, therefore, accepted the allegations in the Notices; that in Notice A alleging that the alterations he undertook [to units 9 and 10] had resulted in a new building on the site". The appellant did not in fact withdraw the appeal under ground (b). He simply did not pursue it in closing submissions because, in the circumstances, it did not help. No adverse inference can be drawn from that. If one strips away the erroneous inference and looks at the underlying facts, what happened to units 9 and 10 – the addition of cladding and internal subdivision – could not amount to the construction of a new building.

[64] In any event, it is submitted that the inspector was wrong to regard the construction of a new building as giving rise to a material change of use in circumstances where it did not by itself create a new planning unit or a new chapter in the planning history. Reliance is placed upon the decision of the Court of Appeal in *Jennings Motors Ltd* v *Secretary of State for the Environment* [1982] JPL 181, in particular the observations of Lord Denning MR, at p183, that "[a]ll that had been done was to erect a new building in place of an old one, on a little portion of the site" and that

the occupiers were entitled to the use of their existing use rights inside the new building.

[65] As to intensification of use, Mr Clay submitted that it had not been open to the inspector to rely upon a material change of use through intensification of use of the site in a case where that was not the change of use alleged in the enforcement notice. In this connection, he cites *Kensington and Chelsea Royal London Borough Council* v *Secretary of State for the Environment* [1981] JPL 50, in which the court held that an inspector had properly rejected an alternative case of material change of use by intensification, since that was not the material change of use alleged in the enforcement notice and was, therefore, not before him on the appeal. Exactly the same reasoning should apply to notice I. A subsidiary point raised in Mr Clay's skeleton argument is that a finding of intensification was not canvassed at the inquiry and it was therefore unfair to rely upon it.

[66] If, instead of taking into account those irrelevant considerations, the inspector had focused upon the question of whether there had been any material change of use in relation to the northern yard, Mr Clay submitted that the only reasonable answer would have been that there had been no such change. The description of activities in notice I, whether in its original or corrected form, is equally apt to describe the situation in the northern yard as the inspector found it to have existed in 1991. There was no material difference in the character of use of the land. Any differences in the number and descriptions of uses on site were immaterial since they all fell within Classes B1, B2 or B8 of the Use Classes Order and, by virtue of section 55(2)(f) of the 1990 Act, uses within the same class cannot give rise to a material change of use. In his skeleton argument, Mr Clay also referred in this connection to article 3(4) of the Use Classes Order, which provides that "where land on a single site… is used for purposes consisting of or including purposes falling within classes B1 and B2…, those classes may be treated as a single class in considering the use of that land" for the purposes of the order.

[67] Mr Clay produced the following written summary of his submissions as to the approach that the inspector ought to have adopted in a case where, as he put it, the area over which unauthorised activities have taken place has not remained consistent: "(1) Does the site on which the activities are taking place contain or include any part of or all of the area on which unauthorised activities were taking place ten years prior to the issue of the [enforcement notice] and have taken place continuously in the intervening period? If yes consider question 2. If no then not immune. (2) Is the use described in the allegation materially different from the description of the use which was taking place on (that part of) the site 10 years prior to the issue of the notice (and continuously in the intervening period). If no then use on site (or relevant part thereof) is immune. If yes consider question 3. (3) If materially different, does the mix of uses fall within the same Use Classes at both beginning and end of the period. If no, then material change of use so no immunity; if yes then no [material change of use], so uses immune."

[68] For the First Secretary of State, Mr Mould started by observing that, having regard to the terms of section 174(2)(d) (appeal on the ground "that, at the date when the notice was issued, no enforcement action could be taken in respect of any breach of planning control which may be constituted by those matters"), the basic issue for the inspector to resolve was whether the mixed use alleged in notice I was no longer capable of enforcement, by virtue of the 10-year time limit.

[69] Mr Mould submitted that the inspector approached the matter in a clear and sensible way, looking at the planning unit as it existed at the date of the enforcement notice and considering whether the mixed use at that date was materially different from the use in 1991. His conclusion that there had been a material change was based upon the combined effect of the various changes. He acknowledged that there were generic similarities in the activities carried on over the period, but considered that the various changes, for example in the way in which they were conducted, the extent to which they were conducted and the buildings within which they were conducted (affecting scale and intensity of use), were such as to give rise to a material change of use. All the matters he took into account were properly related to the character of the use of the land. He made an unremarkable and legitimate planning judgment as a matter of fact and degree.

[70] As to the planning unit, there is no issue over the inspector's conclusions concerning the existing planning units at the site (and his analysis accorded anyway with what was said by Bridge J in *Burdle*), and it is common ground, in particular, that notice I was properly directed to the area of land identified as planning unit C. The inspector's use of that planning unit as a tool for assessing the materiality of any changes during the 10-year period is orthodox and correct: the question is whether the mixed use of that planning unit, at the date of the notice, involved a material change from uses previously carried on during the 10-year period. The fact that there were different planning units at the beginning of the period does not necessarily mean that there has been a material change of use, although it tends to suggest it. In any event, the changes leading to the creation and extent of the present planning unit are matters properly taken into account.

[71] As to the construction of new buildings, Mr Mould submitted that *Jennings Motors* does not assist the appellant. The majority reasoning, in the judgment of Oliver LJ, with which Watkins LJ agreed, was simply that the erection of a new building does not necessarily constitute a new planning unit and give rise to the loss of existing use rights; it is a question of fact and degree in every case: see pp183-184. In the present case, the inspector looked not just at the changes to the buildings themselves but also at the effect this had on the activities carried on within them and in connection with them. Thus, for example, the physical changes have brought about major increases in storage activity and a greater intensity of activity on site. They, in turn, have materially affected the appearance of the site and have resulted in significant increases in traffic movements. These were all relevant considerations that properly informed the overall conclusion that there had been a material change of use.

[72] It is submitted that, having regard to his summary at para 7 of A the decision (where he records the appellant's acceptance that the appeals under ground (b) were not being pursued), the inspector had been entitled to proceed on the basis that the appeal against notice A under ground (b) of section 174 had been withdrawn and that the appellant was therefore to be taken to have accepted that the alterations to units 9 and 10 resulted in a new building. But, in any event, this is said to be an arid point to debate, B since the inspector went on to carry out his own analysis of the nature and effect of the works in question on the basis of the other evidence before him – an analysis that led him to exactly the same conclusion: see paras 80 to 82 and 88 to 89; see also paras 126 to 128.

[73] As to intensification, Mr Mould submitted that it was clear that the inspector had not proceeded upon the basis that there was a material C change of use by intensification, so as to fall into the error, referred to in *Mia Carla*, of relying upon a material change of use different from that identified in the enforcement notice. Intensification, in the form of additional activities, facilities, traffic, etc, is properly taken into account as part of the normal assessment of whether there has been a change in the character of the use of the land so as to give rise to a material change D of use: see, for example, *Lilo Blum* v *Secretary of State for the Environment* [1987] JPL 278, at pp280-281. That is how the inspector proceeded in this case. That is clear both from the detailed analysis and from the conclusions in paras 99 to 100, where it is stated in terms that "[i]t was not just a more intensive use of the site but one where the activities making up the uses, the land area used; the extent of storage, both inside and outside; the E buildings in which the uses were being undertaken and the level of traffic generated, if all taken together, describe a use that is materially different in character and nature".

[74] Mr Mould further submitted that the appellant's reliance upon section 55(2)(f) and the Use Classes Order is misplaced. The Use Classes Order does not apply to mixed use of the land, even if some of the F activities comprised within that mixed use would, if taken by themselves, fall within use classes. In any event, some of the activities comprised within the mixed use are *sui generis* and would not fall within a use class even if taken by themselves. Further, article 3(4) is plainly incapable of assisting the appellant, since it refers to Classes B1 and B2 but not Class B8, yet the activities comprised within the mixed use include storage uses G falling within Class B8, rather than within B1 or B2.

[75] I propose to express my conclusions on this issue relatively briefly, since I accept the substance of Mr Mould's submissions. In my judgment, the inspector had approached the matter correctly, and had reached a carefully reasoned and rational finding that there had been a material change of use. H

[76] The inspector had been right to consider the planning unit and the use as they existed at the date of the enforcement notice, and to consider whether that use was materially the same as at the beginning of the relevant period, or whether there had been a material change of use. No issue is taken over the identification of planning unit C as the relevant

planning unit as at the date of notice I. In my view, it was reasonably open to the inspector to find, on the facts, that the northern and southern yards had constituted separate planning units in 1991 and had subsequently merged into a single planning unit that had also subsequently been extended. These were pre-eminently matters of judgment for him. He did not treat the change in the planning unit as necessarily giving rise to a material change of use, but looked at the change in the planning unit, and the related question of how the site was being used, as part of his overall consideration of whether there had been a material change of use. This was a lawful approach.

[77] So, too, the inspector had not proceeded upon the basis that the construction of a new building, arising out of the works to units 9 and 10, in 1998, necessarily gave rise to a material change of use. Once more, he had considered the relevant works, and their practical consequences for the use made of the site, as part of his overall consideration of material change of use. Such an approach was consistent with what was said in *Jennings Motors* and was again, in my view, lawful.

[78] I agree with Mr Mould that it makes no practical difference whether the inspector had been correct in finding that the ground (b) appeal against notice A had been withdrawn, and that the appellant had thereby impliedly accepted that the works resulted in a new building. The inspector had clearly reached the same finding about the construction of a new building on the basis of his own detailed examination of the evidence concerning the works, and it was a finding reasonably open to him on that evidence.

[79] The inspector had looked at intensification of use as part of a normal assessment of whether there had been a change in the land use character. It had been legitimate for him to do so in determining whether there had been a material change of use to the existing mixed use, as alleged in the notice. He had not found a material change of use by intensification, or departed in any way from the change of use alleged in the notice. Nothing in *Kensington and Chelsea* casts doubt upon the correctness of his approach. There was no unfairness in his taking intensification into account in the way he did.

[80] I do not consider the appellant's case to be assisted by reference to the Use Classes Order. Each of the points made by Mr Mould is, in my view, well founded. I would place particular emphasis upon the threshold point that the Use Classes Order has no application to a mixed use: the mixed use does not itself fall within any class, and a finding of material change of use is not avoided simply by showing that a component falling within a particular class has been substituted for another component falling with the same class. I dealt with a similar point recently in *Belmont Riding Centre* v *First Secretary of State* [2003] EWHC 1895 (Admin) in [31] of the judgment.

[81] The inspector accepted that there were similarities in uses carried on over the relevant period. But he was reasonably entitled to find, as a matter of fact and degree, that the combined effect of the various changes he identified was that the mixed use, at the time of the notice, was

materially different in character and nature from that at the start of the period, and that it had come about through a material change of use. A

[82] I therefore find against the appellant on the material change of use issue.

*(3) Previously established lawful use issue*

[83] I have already mentioned how the inspector dealt with the related section 195 appeal, concerning the application for a certificate of lawful development, including his finding at paras 104 to 105 that only a very limited number of uses could be regarded as lawful, and should be excluded from the scope of the requirement in notice I to cease activities on the site. The issue here is whether he had erred in his consideration of whether other commercial uses had been carried on by the appellant long enough to become lawful. B C

[84] The relevant passage of the decision reads in full:

> 104. The Council's planning witness accepted that the use of the site for hay and straw dealing and agricultural contracting was lawful; it had gained that lawfulness in July 1992 having commenced 10 years before that date. The appellant also only referred to these two uses as being lawful, both in questions to his own planning witness and the Council's planning witness. These were the only alleged uses that the Council considered should not be required to cease other than agriculture which it was agreed was not "development". D
>
> 105. No similar point was put regarding the other uses that the appellant claimed to have carried out himself at the site and no detailed evidence was put forward regarding these other uses; that contained in the supplementary proof of the appellant's planning witness was really only diagrammatic in relation to where it took place and general in terms of the type of use taking place in a building or amongst the open storage. The appellant was also just one of a number of people/companies that carried out such uses which were all part of the mixed use which I have already concluded materially changed in the relevant 10 year period. E F

[85] Mr Clay submitted that the inspector had erred in law in considering only the past 10 years, and had acted irrationally in dismissing the appellant's case as to other lawful uses. As to the former point, it is not enough to compare the existing use with the use 10 years ago, since immunity from enforcement may have arisen by reason of some previous period of continuous use. As the inspector said in para 104, in relation to hay and straw dealing and agricultural contracting, "it had gained that lawfulness in July 1992 having commenced 10 years before that date". Yet, in relation to other activities, he had not asked himself whether there was a continuous period of 10 years within which they had been carried on, and the last sentence of para 105 shows that he considered only the past 10 years. G H

[86] The point on irrationality relates to the first sentence of para 105, which is said to be a travesty of the evidence put forward. In addition to a detailed schedule from the appellant's planning expert, there were 45 or so statements from people who occupied the land at various times, giving primary evidence of the activities carried on there. The inspector does

not refer here to any of them. In the light of that body of evidence, it was perverse to deal with the matter in this way.

[87] Mr Mould submitted that the inspector had made it clear what he was deciding and why. It was common ground, and found by him, that hay and straw dealing and agricultural contracting had become lawful. But no similar point was put forward as regards other uses. There was certainly evidence that activities were being carried on as part of a mixed use of the site, and the inspector's earlier detailed analysis of the situation up to 1991 shows that he had had this evidence clearly in mind. It was not, however, contended that any specific individual component of that mixed use had become lawful in its own right. The appellant's closing submissions at the inquiry did not raise the point, but focused upon the period from 1991. Even when arguing under ground (f) that it would be excessive to uphold all the requirements of the notices, it was not contended that there existed a fall-back position of previously established lawful use in respect of any specific activities. Accordingly, what the inspector says is an accurate reflection of the way the case was put to him, and he dealt with the matter sufficiently in the circumstances.

[88] On this issue, too, I accept Mr Mould's submissions. It is unsurprising that the inspector had found that there had been at least 10 years' continuous use of the site for hay and straw dealing and agricultural contracting, since the point was common ground between the parties. He had been entitled, however, to limit his findings to that point. He seems to me to have been justified in his view that no corresponding case had been advanced in relation to any other specific use. The 45 or so witness statements were, ultimately, relied upon by the appellant in relation to the general issue of absence of material change of use, rather than in support of any such specific case, and the inspector's handling of the issue of material change of use shows that he had them clearly in mind. It was not unreasonable for him to conclude that the evidence of the appellant's planning expert, which, again, he had clearly had in mind, did not amount to a case of previously established lawful use in respect of any specific uses. The matter was simply not argued in that way in the appellant's closing submissions. In the circumstances the inspector cannot fairly be criticised for having dealt with the matter as he had.

[89] I should record for completeness that, in my view, the inspector had plainly appreciated that the 10-year period for determining whether a lawful use had been established was not limited to the 10 years immediately preceding the application. That is clear from para 104 of his decision. The last sentence of para 105 does not disclose any different or mistaken understanding of the test.

*(4) Notice A appeal issue*

[90] Notice A related to the works carried out to units 9 and 10 in 1998. The breach alleged was "the addition of block work and cladding to, and the re-roofing of, former open sided agricultural barns resulting in the new building (Units 9 and 10) shown … on the plan". This was said to be operational development carried out without planning permission. The inspector dismissed the ground (a) appeal against notice A, finding

that the works constituted inappropriate development in the green belt, A and that there were no very special circumstances that outweighed the presumption against such development.

[91] Mr Clay submitted, first, that if the appellant succeeds on the second-bite issue or the material change of use issue, it would be right to remit the notice A appeal to the Secretary of State at the same time, in order to permit him to reconsider the ground (a) issue concerning B operational development. I do not see why that should follow, but, having decided that, subject to any further submissions on discretion, the appeal should succeed on the second-bite issue, I will, if necessary, hear further submissions as to the consequences for the notice A appeal.

[92] Mr Clay submitted that, in any event, the inspector had erred in two respects in reaching his conclusion on the notice A appeal. C

[93] The first point concerns the question of whether the works were permitted development within the Town and Country Planning (General Permitted Development) Order 1995, or would have been permitted development but for one technical point concerning the proximity of the land to an aerodrome. The inspector stated, in para 119:

D

> 119. There was some confusion about whether or not the building and its extension could have been erected as "permitted development" for agricultural purposes and although the Council may have wrongly told the appellant this was the case at some time in the past, it was agreed that the buildings could not be permitted development due to the proximity of the site to an aerodrome and the size of the building involved. The appellant admitted in his evidence that it was formed from second hand industrial E rather than agricultural buildings but I agree that currently there is little difference, with the same profiled steel sheeting often used for the cladding and roofing on both.

[94] It is submitted that he should have made a positive finding in the appellant's favour. It is not in issue that agricultural use of the land is lawful. On that basis, the inspector should have considered whether F the works were permitted development under Part 6 of Schedule 2 to the 1995 Order, where permitted development under Class A is defined as: "The carrying out on agricultural land comprised in an agricultural unit of 5 hectares or more in area of – (a) works for the erection, extension or alteration of a building… which are reasonably necessary for the purposes of agriculture within that unit". It is true that para A1(e) provides that G development is not permitted by Class A if "the height of any part of a building, structure or works within 3 kilometres of the perimeter of an aerodrome would exceed 3 metres", which was the case here. But the inspector ought to have considered whether, but for that point, it would have been permitted development and, if so, whether the proximity to the aerodrome was, in practice, a good reason for refusing H planning permission.

[95] I accept Mr Mould's submissions that such reliance upon the 1995 Order is unsustainable for several reasons. First, the land was not in agricultural use at the relevant time, but was in mixed use, including the various commercial activities found by the inspector. The provisions of Part

6 of Schedule 2 to the 1995 Order, which relate specifically to agricultural buildings and operations, had no application to land in such mixed use. Second, it was not contended before the inspector, and there was no evidence before him, that the works were "reasonably necessary for the purposes of agriculture" within the unit. Third, once it was accepted that, because of the proximity to the aerodrome, the works could not amount to permitted development, it was not necessary for the inspector to consider the matter in terms of the 1995 Order. Fourth, there was, unsurprisingly, no suggestion before the inspector that the building should be allowed to remain because of a reasonable need for it as an agricultural building: see, in particular, para 153.

[96] The second point raised by Mr Clay, although it was not in the claim form or skeleton argument, is that the inspector had applied the wrong test under section 55(2)(a). That provision excludes from the scope of development the carrying out of works of alteration etc that: (i) affect only the interior of the building; *or* (ii) do not materially affect the external appearance of the building. That is a *disjunctive* test. In para 125, however, the inspector expressed it as a *conjunctive* test, using "and" instead of "or". He said:

> ... I agree that the definition states that it is not development if works for the maintenance, improvement or other alteration of a building are works which firstly, affect only the interior of the building and secondly, do not materially effect [sic] the external appearance of the building.

[97] Again, I accept Mr Mould's submissions in response to that point. Even if the inspector had, technically, been in error in referring to the statutory test in conjunctive rather than disjunctive terms, it gets the appellant nowhere on the facts, since the inspector went on to explain, in para 126, why, in his view, the works did not affect only the interior of the building and did materially affect the external appearance of the building. Thus, neither limb of section 55(2)(a) had been satisfied and nothing turns on whether a disjunctive or conjunctive approach was applied.

[98] Subject, therefore, to the implications of the appellant's success on the second-bite issue, the appellant's case in relation to the notice A appeal fails.

*(5) Deemed planning permission issue*

[99] In his decision, the inspector rejected an argument by the appellant that the combined effect of section 173(11) and an enforcement notice issued in 1996 had been to confer deemed planning permission in respect of the uses that the 1996 notice did not require to cease.

[100] I have already set out the terms of section 173(11). It applies where an enforcement notice in respect of a breach of planning control "could have required any buildings or works to be removed or activity to cease, but does not do so" and all the requirements of the notice have been complied with. In those circumstances, then "so far as the notice did not so require, planning permission shall be treated as having been granted... in respect of development consisting of the construction of the buildings or works or, as the case may be, the carrying out of the activities".

[101] The 1996 notice was expressed to relate to the whole site. The breach of planning control alleged was: "Without planning permission, change of use of the said land from use for agriculture to a mixed use for agriculture and an engineering contractors' depot involving the storage of plant, equipment, materials, vehicles, portable buildings and machinery". What the appellant was required by the notice to do was: "(i) Cease the use of the land for an engineering contractors' depot... (ii) Remove from the land all plant, equipment, materials, vehicles, portable buildings and machinery used or stored in connection with the use of the land as an engineering contractors' depot..."

[102] The notice was appealed on a number of grounds. At the inquiry, it was made clear, by counsel for the council, that the notice was concerned only with the use of the site for what has been described as "the McNicholas use", namely use as a depot for vehicles and equipment during cable-laying operations in the area by a contractor called McNicholas. There were various adjournments, and indications were given of an intention by the council to amend the notice to cover other operations taking place on the site. In the event, however, the appeal was withdrawn for a number of reasons, the principal of which was the council's assurance that the notice was directed only at the McNicholas use, which had in fact long ceased.

[103] The submission made by Mr Clay on the basis of those facts is that the effect of section 173(11) was to confer a deemed planning permission on all other uses of the site existing at the time of the 1996 notice.

[104] The first and, in my judgment, insuperable hurdle faced by that submission is the decision of the Court of Appeal in *Scott* v *Secretary of State for the Environment, Transport and the Regions (Enforcement Notice)* unreported 16 October 2000, on appeal from the first instance judgment reported at [2000] JPL 833. That decision makes clear that section 173(11) applies only to uses of the land that are alleged by the enforcement notice itself to be in breach of planning control. If a use is not alleged to be in breach of planning control, the notice cannot require the use to cease, and deemed planning permission cannot arise out of the failure of the notice to require the cessation of that use. The provision therefore has no application to uses that are not alleged by the notice to be in breach of planning control. As it was put in para 15 of the judgment of Schiemann LJ:

> The enforcement notice manifestly did not allege that construction of the culvert had been carried out in breach of planning control. In those circumstances the enforcement notice could not have required the removal of the culvert. In those circumstances section 173(11) is not in play and there is no question of any deemed planning permission for the culvert...

[105] Mr Clay seeks to bring the appellant's case within the scope of the reasoning in *Scott*, by contending that the breach of planning control alleged in the 1996 notice was the entirety of the mixed use of the land, the true nature of which was known to the council at the time, but that the council chose, by the notice, to require cessation only of the McNicholas use. Accordingly, he submitted, there was under-enforcement of a kind to

which section 173(11) applies and the balance of the mixed-use benefits from deemed permission.

[106] In my view, the 1996 notice is not to be construed in that way. The "mixed use" alleged by the notice to be in breach of planning control was limited to agricultural use plus the McNicholas use, rather than the entirety of the activities on site (and as comprised, for example, within the mixed use alleged in notice I), and there was a direct correspondence between the breach alleged and the requirement to cease the McNicholas use. That was also the basis upon which the parties proceeded. The council made clear that the notice was directed only at the McNicholas use (there being no objection to agricultural use), and that was one of the reasons why the appeal against the notice had been withdrawn. Even if there was a vagueness or ambiguity in the wording of the notice, the fact is that the parties proceeded on the basis of a narrow construction, and the present submission, that the notice was directed to the entirety of the activities on site and that everything but the McNicholas use should therefore gain the windfall benefit of deemed planning permission, runs counter to the position accepted in practice by the appellant at the time.

[107] In his skeleton argument, Mr Clay also relied heavily upon para 2.10 of Circular 10/97, which deals with section 173(11). Some of the language in the paragraph suggests that deemed planning permission is conferred by section 173(11) on "all the remaining uses or activities" on the land save those the cessation of which is required by the enforcement notice. This could be taken to indicate a broader construction of the provision than that adopted in *Scott*. Mr Mould accepted that the paragraph was confusing or ambiguous in part, but rightly submitted that the circular cannot affect the construction of section 173(11) or detract from the effect of the decision in *Scott*. No doubt, consideration will be given to improving the text of the circular so as to reflect the decision in *Scott* more clearly.

[108] Those observations on the circular do not affect my rejection of the appellant's case on the deemed planning permission issue.

*Overall conclusion*

[109] For the reasons given, I find in favour of the appellant on the second-bite issue alone. As indicated, I will hear from counsel on questions of relief.

*Appeal allowed.*

# R (on the application of Mount Cook Land Ltd and another) *v* Westminster City Council

[2003] EWCA Civ 1346

COURT OF APPEAL
*Auld, Clarke and Jonathan Parker LJJ*
14 October 2003

**Planning application – Material considerations – Whether** A
**alternative proposals for planning permission material consideration – Costs – CPR Part 54 guidance on award of costs of permission stage**

R, the owner of a 999-year lease of premises in Oxford Street, London W1, applied for planning permission to redevelop the building. MC, the B
freeholder, wished to acquire the lease in order to develop the property along with adjoining land that it also owned. It objected to R's application, and advanced its own, alternative, development proposals. In May 2002, the respondent council granted conditional planning permission to R. MC's application for permission to apply for judicial review of that decision was rejected by Ouseley J on the ground, *inter alia*, that the council were entitled to disregard MC's alternative proposals C
because it would have been irrational for the council to have relied upon these as a basis for refusing permission to R's scheme. MC renewed its application for permission to claim for judicial review. Moses J also took a similar view to that of Ouseley J and rejected its claim, emphasising that an applicant for planning permission whose proposal conforms to local planning policies does not have to show planning benefit, D
but simply a lack of planning harm. MC appealed, contending, in effect, that its alternative proposals were a material consideration and that the council were entitled to refuse permission on the ground that a grant would make a future change of use that would be contrary to their local planning policy materially harder to resist. The matter of R's proposed internal alterations was relevant because these would prejudice MC's E
alternative proposals.

***Held*** The appeal was dismissed and the claim for judicial review was rejected.

The critical question, one of mixed law and fact, was whether the existence of a possible alternative scheme, one that would be F
more beneficial, in planning terms, than that proposed in a planning application was a "material consideration" for the purpose of deciding a planning application: see [21]. It would be highly harmful to the efficient and otherwise beneficial working of the system of planning control if decision makers were required to consider possible alternatives, of which, on the facts before them, there was no likelihood or real possibility of being introduced in the foreseeable future: see [35]. The approaches of Ouseley and Moses JJ were correct. The alternative proposals of

MC were not material considerations within the meaning of sections 54A or 70(2) of the Town and Country Planning Act 1990, or, if they were, they were of such negligible weight that the court would be entitled to refuse permission to apply for judicial review because the council could not reasonably have taken any notice of them: see [36]; *R (on the application of Scott)* v *North Warwickshire Borough Council* [2001] EWCA Civ 315; [2001] 2 PLR 59 considered. The council had been entitled to reject MC's contention that the internal works proposed by R, and which were outside planning control, were relevant to a complaint of "planning creep" and would prejudice MC's proposals: see [44]. With regard to the award of costs at the permission stage of an application for judicial review generally, and in the absence of exceptional circumstances, the costs of, and occasioned by, attendance of a party at a permission hearing should not be awarded against a claimant. But in respect of the costs of filing an acknowledgment, where a claimant is ultimately unsuccessful in a claim for judicial review, the court should exercise a broad discretion.

**Cases referred to in the judgment**

*Berkeley* v *Secretary of State for the Environment, Transport and the Regions (No 1)* [2001] 2 AC 603; [2000] 3 WLR 420; [2000] 3 All ER 897; (2001) 81 P&CR 35; [2000] 3 PLR 111; [2001] JPL 58; [2000] EGCS 86, HL

*Bolton Metropolitan Borough Council* v *Secretary of State for the Environment* (1991) 61 P&CR 343; [1991] JPL 241, CA

*Jolly* v *Jay* [2002] EWCA Civ 277; The Times 3 April 2002

*Lake District Special Planning Board* v *Secretary of State for the Environment* (1984) 47 P&CR 157; [1981] JPL 363, QB

*Leach* v *Commissioner for Local Administration; R (on the application of Leach)* v *Commissioner for Local Administration* [2001] EWHC Admin 455; [2001] CP Rep. 97; [2001] 4 PLR 28

*New Forest District Council* v *Secretary of State for the Environment* (1996) 71 P&CR 189; [1996] JPL 935, QBD

*Nottinghamshire County Council* v *Secretary of State for the Environment, Transport and the Regions* [2001] EWHC Admin 293; [2002] 1 P&CR 30; [2002] JPL 161 QB

*R (on the application of Scott)* v *North Warwickshire Borough Council* [2001] EWCA Civ 315; [2001] 2 PLR 59; [2001] PLCR 31

*R* v *Honourable Society of the Middle Temple, ex parte Bullock* [1996] ELR 349; [1996] COD 376

*R* v *Secretary of State for Wales, ex parte Rozhon* (1993) 91 LGR 667; [1994] COD 111; [1993] NPC 72

*Redevco Properties* v *Mount Cook Land Ltd* [2002] All ER (D) 26

*Somak Travel Ltd* v *Secretary of State for the Environment* (1988) 55 P&CR 250; [1987] JPL 630, QBD

*South Buckinghamshire District Council* v *Secretary of State for the Environment, Transport and the Regions* [1999] PLCR 72; [1998] NPC 106

*South Lakeland District Council* v *Secretary of State for the Environment* [1992] 2 AC 141; [1992] 2 WLR 204; [1992] 1 All ER 573; (1992) 90 LGR 201; 64 P&CR 128; [1992] 1 PLR 143, HL

*Stringer* v *Minister of Housing and Local Government* [1970] 1 WLR 1281; [1971] 1 All ER 65; (1970) 68 LGR 788; 22 P&CR 255; [1971] JPL 114

*Trusthouse Forte Hotels Ltd* v *Secretary of State for the Environment* (1986) 53 P&CR 293; [1986] 2 EGLR 185; 279 EG 680; [1986] JPL 834

**Appeal against a decision of Moses J**

This was an appeal by the appellant, Mount Cook Land Ltd, against a decision of Moses J refusing an application for judicial review of a grant of planning permission by the respondents, Westminster City Council, to an interest party, Redevco Properties.

*John Steel QC, Robert White and Stephen Whale (instructed by Stephenson Harwood) appeared for the appellants, Mount Cook Land Ltd and Mount Eden Land Ltd.*

*Timothy Corner QC and Robert Palmer (instructed by the solicitor to Westminster City Council) represented the respondent council.*

*The following judgment was delivered.*

**AULD LJ:**

[1] This is a claim by the appellant (Mount Cook) for judicial review of the grant by the respondents (the council) to the interested party, Redevco Properties (Redevco) of "operational" planning permission for external alterations to a building (formerly housing the C&A store) at 200-212 Oxford Street, London W1 (the building). The matter originally came before the court by way of a renewed oral application by Mount Cook for permission to appeal a refusal by Moses J, on 26 September 2002, of its renewed application to him for permission to apply for judicial review. On 27 November 2002, this court granted permission to appeal and to claim judicial review, and retained the claim for its own decision.

*Facts*

[2] The building is within the East Marylebone Conservation Area, as designated by the council, the local planning authority. The statutory development plan for the area is the Westminster unitary development plan, adopted by the council in July 1997. The relevant policies are SS2 and DES7, the former "to protect retail floorspace in large stores trading on several floors" inside the central zone and the latter "to preserve or enhance the character or appearance of... [the council's] conservation areas". It is also within an area for which the local plan policy is to protect department stores, a policy that Mount Cook claims, in these proceedings, to support.

[3] The building fronts onto the north side of Oxford Street and backs onto Market Place, a series of short highways in an H formation, which, until recently, had had little attraction for members of the public. As a result of an initiative in the late 1990s, known as "the Oxford Market Initiative" (OMI), by two major local landowners, supported by Mount Cook and the council, parts of Market Place have been improved and made more attractive to the public. However, the improvements did not extend to that part of Market Place immediately to the rear of the building, which remains a rather rundown area by the standards of this part of the West End.

[4] Mount Cook is the freeholder of the building. Redevco has a 999-year lease of it, at an annual rent of £1,139, not expiring until 2912,

A that is, not for another 910 years. Redevco's interest in the building is thus close to that of a freeholder, a strong position that has clearly exercised Mount Cook, which would like to acquire the leasehold interest to enable it to develop the building for its own purposes. It has substantial property interests to the rear and north of the building, all of which, including the building, are known as the Langham Estate, which it manages through a B managing agent. In all, the estate comprises more than 400 tenancies, some long-term, some short term, most of which are in commercial uses, mainly retail in the form of shops, restaurants, and offices. Relevant to Mount Cook's aspirations, some of the leases in respect of buildings close to the rear of the building are for terms that will expire in the near future.

[5] It is common ground that Mount Cook has sought to bring pressure C on Redevco to yield to its development ambitions by reliance upon its entitlement, under the lease, to refuse consent to alterations to the building that Redevco sought to make, and by objecting to various applications by Redevco for planning permission. As to the former, Redevco succeeded, on 3 December 2002, in obtaining from Mr Paul Morgan QC, sitting as a deputy judge of the High Court, declarations that Mount Cook's refusals D of consent were unreasonable and that Redevco was entitled to make the alterations without its consent; see *Redevco Properties* v *Mount Cook Land Ltd* 3 December 2002, at [19](7) and (10), [38], [40] and [45]. On 20 February 2003, this court refused Mount Cook permission to appeal that decision. On the planning front, Mount Cook has objected to at least three planning applications in respect of the building. The first was for E change of use of the upper floors from retail to office and residential uses. It was withdrawn, after some revisions, in May 2002. The second, which was made in revised form in February 2002, was to make various alterations. It was granted in May 2002, and is the subject of this appeal. And the third was made in around May 2002, following the withdrawn first application to which it was similar, namely for change of use of some F of the upper floors from retail to office or residential use. The council have yet to decide on that application.

[6] The planning permission under challenge was for relatively minor physical external alterations, involving: (i) the installation of new shopfronts and canopies to the ground floor of each of its three street elevations, including that looking onto the southern part of Market Place; G (ii) the partial infill of a lightwell; (iii) the installation of glass louvres on the second floor and of bronze louvres on the ground floor; and (iv) the replacement of black-painted windows with clear windows on the first and second floors. The proposed alterations to the Market Street elevation, although relatively minor, would be a distinct enhancement of the character of the building, and of that part of Market Place onto which H it looks.

[7] Mount Cook objected to Redevco's application, as it had objected to an earlier form of it, on the grounds that Redevco had not considered how its proposals would preserve or enhance the character and appearance of the conservation area, and that some of the alterations might prejudice the future use of the upper floors of the building for retail purposes.

In the meantime, in early 2002, Mount Cook had commissioned the production of a number of design options for the improvement of the southern part of Market Place to the immediate rear of the building, with a view to enlivening the area by providing more retail frontages. In March 2002, it put them to the council as a logical extension of the scheme of improvements already achieved by the OMI in the northern section of Market Place. Its proposals were supported by two other major local landowners, one of which had been jointly responsible for that initiative. It contrasted them with Redevco's proposed alterations to the rear of the building, which, it claimed, would "nullify" Mount Cook's proposals. However, it did not embody them in any planning application of its own, indicating that it would not do so unless agreement could be reached for Redevco's proposals to conform broadly with its own.

[8] The council's response to Mount Cook's alternative proposals was that they were too vague for proper consideration and advice by their officers, although they proffered some "initial" advice indicating that the proposals were likely to be unacceptable. However, the council informed Mount Cook that its objections to Redevco's application would be put before the planning applications subcommittee meeting on 21 March 2002, when they were due to consider the application. (The council later accepted that this initial advice had been based upon the wrong set of plans.)

[9] The council's director of planning and transportation, in a report prepared for the subcommittee, recommended the grant of conditional permission. He referred to Mount Cook's main objection in general terms, and identified, as background papers available to the members of the subcommittee, various correspondence from Mount Cook's advisers setting out its objections in detail, in particular by comparison with its own proposals. He advised the committee, in the following terms, that those proposals were irrelevant to the determination of Redevco's scheme:

> An alternative scheme has been submitted on behalf of the freeholders for new shop fronts to the Market Place elevation of the building. Given that each case is treated on its own merits, these proposals are not considered relevant to an evaluation of this application.

In subsequent correspondence, the council stated that the planning officer responsible had circulated the Mount Cook correspondence to the members of the subcommittee, who had fully considered it. Mount Cook's solicitor's note of the meeting records that the planning officer responsible for the application had referred to Mount Cook's advisers' letters, summarised its concerns as set out in the report for the meeting, "with the addition of a reference to improving Market Place", and advised that, as to the complaint about prejudicing future applications, Redevco's application had to be considered on its merits. The note also records that he and another officer expressed the view that the proposal would enhance the building and the appearance of the conservation area. Notwithstanding Mount Cook's objections, the subcommittee approved the application.

A [10] There then followed further correspondence in which Mount Cook
complained to the council about the decision, repeated its earlier assertion
that Redevco's proposed alterations would undermine the likelihood
of any extension of the OMI to the southern part of Market Place, and
threatened not to pursue its proposals for that extension if Redevco's
approved alterations went ahead. At the same time, Mount Cook's
B solicitor wrote to the council, informing them of Mount Cook's intention,
if necessary, to seek judicial review of the subcommittee's decision on a
number of likely grounds, including:

> failure to have regard or proper regard to the effect that implementation
> of the approved proposal is likely to have on the conservation area within
> which the building is situated particularly with regard to its adverse impact
> C upon the prospects of completing improvements to the conservation area
> begun by the Oxford Market Initiative.

[11] The council, in a letter in reply, indicating their intention to refer
back for clarification to the committee another objection of Mount Cook
immaterial to this appeal, stated that they had had proper regard to the
D impact that the proposal was likely to have upon the conservation area.
Mount Cook, after sight of a report by the planning officer for the further
committee hearing, decided not to pursue the objection giving rise to it.
In that second report, the planning officer referred again to Mount Cook's
main ground of objection to the application:

> Since this application was presented to the Sub-Committee in March there
> E has also been further correspondence on behalf of the freeholder. This refers
> to the adverse implications of the proposed works, in particular the creation
> of dead frontage to north elevation of 200 Oxford Street, for potential
> improvements in Market Place linked to extending the "Oxford Market
> Initiative". Sub-Committee considered a similar objection previously. There
> is no agreed package of environmental improvements for this part of the
> public highway…

F

In the result, the council, by letter of 9 May 2002, formally granted
conditional permission to Redevco, in accordance with its application,
and Mount Cook duly proceeded with its application for permission to
claim judicial review.

[12] As Moses J observed in [9] of his judgment, there is no doubt that
G the subcommittee had been advised that Mount Cook's proposals were
irrelevant to their consideration of Redevco's application. In neither of
the planning officer's reports was it suggested that those proposals were
matters that they ought to take into consideration.

*Applications at first instance for permission to claim judicial review*

H [13] Mount Cook's application was first considered and rejected as
unarguable on paper by Ouseley J. In his observations, he said that
the council were entitled to disregard Mount Cook's alternative
proposals because it would have been irrational for the council to have
relied upon them as a basis for refusing permission to Redevco's beneficial
scheme. He added that even if Mount Cook's scheme could have been

regarded as more beneficial than that of Redevco, there was no evidence A
that it would be capable of implementation, or that a refusal of permission would assist their implementation. Accordingly, the difference in practice between ignoring Mount Cook's scheme and giving it no or negligible weight, as the advice to the council would obviously have been, was nil.

[14] Mount Cook indicated its intention to renew its application by way B
of oral hearing. The council, having filed an acknowledgment of service, pursuant to the new provision for it in CPR 54.8, indicating their wish to take part in the judicial review and to contest the claim, also indicated their intention to attend the renewal hearing. The hearing before Moses J lasted nearly a full day, and was, for all practical purposes, a hearing of the substantive claim. Both parties appeared by counsel, who had provided C
the court with skeleton arguments, and both parties filed all the evidence that would have been required at a substantive hearing, had it gone that far.

[15] Moses J took a similar view to that of Ouseley J, emphasising also that an applicant for planning permission, whose proposal conforms to local planning policies, does not have to show planning benefit but D
simply lack of planning harm. On the main issue as to the relevance to the planning decision of Mount Cook's alternative proposals, he said, in [12], [13] and [17] of his judgment:

> [12]... The Council had no power to refuse permission on the basis that the proposal would not enhance he character of the area; see *South Lakeland District Council* v *Secretary of State for the Environment* [1992] 1 All E
> ER at page 573. The only obligation of the Council was to consider whether the development left the character or appearance of the conservation area unharmed. In fact, the officer advised that the proposals would enhance the character and appearance of the building and the conservation area, and meet the policy tests... The fact that there might have been a better scheme for enhancement was, in the view of the Council, nothing to the point. That, in my judgment, was an approach that the Council was perfectly entitled F
> to adopt.
>
> [13] Further, the suggestion that the Council was bound to consider the alternative scheme is, in my view, fallacious because there was no basis for suggesting that there was any possibility of the proposals of the claimants coming to fruition... Redevco owns a 999 year lease of number 2000. It is plain that they were intent upon developing the north side of the building in the way they proposed. The claimants had no power whatsoever to compel G
> them to do otherwise. If permission were refused then the northern aspect of that building would remain as it was. In my view, it is not arguable that it is open to the claimants to seek to exercise some control over the building in the face of Redevco's leasehold interest, by saying that as a matter of law the Council was under an obligation to consider alternative schemes as disclosing a better proposal.
>
> ... H
>
> [17] There is, in my view, nothing in ground 1. The claimants had no realistic prospect of being able to force Redevco to adopt their plan. The Council was correct in law to regard the existence of rival proposals as being irrelevant. Even it is was not irrelevant, it was bound to make no difference to the result, as Ouseley J said when refusing permission in writing.

A [16] Moses J also rejected, as irrelevant and unarguable, a second objection of Mount Cook, that the grant of permission would be likely to prejudice, contrary to the council's planning policy SS2, the future retention of the upper floors of the building in retail use.

[17] In the result, he refused permission to apply for judicial review and ordered the costs of the application, which he summarily assessed at
B £11,508.13, to be paid by Mount Cook to the council.

*Issues*

[18] There are four issues in this judicial review claim that the court, in granting permission to appeal from Moses J's refusal of permission, directed it should retain and determine for itself. Two are matters of
C planning law, the third concerns a court's discretionary power to refuse relief, and the fourth is one of costs at the permission stage of claims for judicial review. They are:

(1) whether a local planning authority, when determining a planning application, are entitled to grant it without regard to a possibility, where drawn to their attention, of an alternative and preferable proposal;

D (2) whether a local planning authority, in determining a planning application for operational development of land, are entitled to refuse it on the ground that a grant of it would make a future change of use contrary to their local planning policy materially harder to resist;

(3) whether, in the event of the court holding that the council had erred in law in granting Redevco's application for planning permission, it would
E nevertheless be entitled to dismiss Mount Cook's claim for judicial review in the exercise of its discretion; and

(4) the circumstances in which a court, on an oral application for permission to claim judicial review, may award costs to a defendant who has attended and successfully resisted the application.

*Issue 1: Materiality to a planning application of a possibility of an alternative*
F *preferable proposal*

[19] As I have indicated, Ouseley and Moses JJ were of the view that the council were entitled to disregard Mount Cook's alternative proposals, notwithstanding its representations to it that they were preferable in planning terms to those of Redevco. Mr John Steel QC, for Mount Cook, in challenging Moses J's ruling, submitted that he had
G been wrong in law in holding that its alternative proposals for the building and for the southern part of Market Place were irrelevant to the council's determination of Redevco's application for operational development to the northern elevation of the building. In particular, he argued that the judge had erred in ruling that, since the council had considered Redevco's application acceptable in
H planning terms, they had no power to refuse it on the basis that there was, or might be, some other preferable scheme. He also criticised the judge's reliance upon Mount Cook's lack of effective control of the building so as to give effect to its proposals as further support, if necessary, for his view of the immateriality, or lack of weight, of its suggested alternative.

[20] The starting points, as Mr Steel noted, are sections 54A and 70(2) of the Town and Country Planning Act 1990, which provide:

> 54A Where, in making any determination under the planning acts, regard is to be had to the development plan, the determination shall be made in accordance with the plan unless material considerations indicate otherwise.
>
> 70(2) [in making a determination of a planning] application the [local planning authority] shall have regard to the provisions of the development plan, so far as material to the application, and to any other material considerations.

[21] The critical question, which, it seems to me, is one of mixed law and fact, is, therefore, whether the existence of a possible alternative scheme is more beneficial in planning terms than that proposed in a planning application is a "material consideration" for this purpose. The 1990 Act gives no help as to what may constitute such a consideration, but the following words of Cooke J in *Stringer* v *Minister of Housing and Local Government* [1971] 1 All ER 65, at p77b, are usually taken as an all-context starting point:

> any consideration which relates to the use and development of land is capable of being a planning consideration. Whether a particular consideration falling within that broad class is material in any given case will depend on the circumstances.

*Submissions*

[22] Mr Steel submitted that Cooke J's broad description includes the consideration, in this case, of the existence of a possible alternative, preferable use for the land in question. He maintained that it would be open to a decision maker to refuse planning permission for an otherwise acceptable development where he considers it desirable, in planning terms, to preserve an alternative option, since to grant permission in such circumstances would, or could, amount to a wider planning harm. In that sense, he defined "planning harm" as an absence of planning benefit; for example, a failure in the application proposal to assist some general local planning policy. He sought support for that submission in the judgment of Mr Christopher Lockhart-Mummery QC, sitting as a deputy High Court judge, in *Nottinghamshire County Council* v *Secretary of State for the Environment, Transport and the Regions* [2001] EWHC 293 (Admin); [2002] 1 P&CR 389, a case in which the deputy judge upheld the decision of an inspector refusing permission for a residential development in the light of his view that it was desirable to preserve an option of retaining the land in question for educational use. The deputy judge dealt with the relevance of such a consideration as a matter of principle at [36] of his judgment:

> If the judgment is made, whether through the development plan process or indeed outside it, that it appears desirable to preserve the option of using a piece of land for a purpose seen to be of benefit in the public interest for the country or the local community, this is, in principle, a material planning consideration for the purposes of sections 70(2) and 54A of the Act. I understood this to be common ground in the case. The weight given to that consideration will vary hugely from case to case... Each case will turn on

> A its own merits, but the importance of the project or proposal, its desirability in the public interest, are undoubtedly matters to be weighed. Therefore, in considering whether to grant planning permission for a proposal (use B) which will pre-empt the possibility of desirable future use (use A), the relative desirability of the two users have to be weighed. *In striking the balance, the likelihood of use A actually coming about is doubtless a highly material consideration.*

B

(Judge's emphasis.)

[23] As can be seen from those observations, the deputy judge had in mind, not only the materiality of such a consideration, but also its weight. He included in the latter the importance and desirability of the alternative proposal to the public interest, and seemingly, in his
C reference to "striking the balance", the likelihood of it coming about. In practice, at the margins, the questions as to whether a consideration is material, and, if so, what, if any, weight should be given to it, shade into each other.

[24] But Mr Steel sought to draw two further propositions from *Nottinghamshire County Council* of particular relevance to the facts of
D this case.

[25] First, he submitted that it is open to a decision maker to refuse permission because of the existence of an alternative, possibly more desirable, scheme in planning terms, even where there was no evidence upon which he could conclude that it is likely to go ahead. That was certainly part of the reasoning of the deputy judge in the following
E passage, in [40] of his judgment:

> I accordingly hold that, subject to matters to which I turn below, it was in principle open to the Inspector to refuse residential development in the instant case, in the light of his conclusion that it was desirable to preserve the option of retaining the appeal site for educational use, albeit that he made no finding that it was more likely than not that the site would effectively be
> F put to educational use.

Such a proposition, which the deputy judge derived in part from the judgment of Mr George Bartlett QC, sitting as a deputy High Court judge in *South Buckinghamshire District Council* v *Secretary of State for the Environment, Transport and the Regions* [1999] PLCR 72, at pp79-80, logically follows from his first proposition (in [22] above) in those cases
G in which the importance or desirability in the public interest of preserving a particular alternative option is so great that the decision maker could reasonably conclude that to grant the application in the circumstances would, or could, constitute a planning harm. Again, both the preferability in planning terms of an alternative scheme and the degree of possibility or likelihood of it coming about may, depending upon their relative strength,
H go to materiality or to weight. In the context of this case, Mr Steel placed much emphasis upon: (i) the claimed public importance and desirability of extending, in the outline way in which Mount Cook proposed, the OMI, including the provision, in the rear elevation, of the building of *both* retail displays and access to the retail areas behind them from Market Place; and (ii) the fact that Mount Cook had alerted the council to its proposals,

and also to its contention that grant of permission to Redevco to make its alterations would preclude the possibility of them coming about.

[26] However, Mr Steel needed, in the circumstances of this case, to advance a second argument, namely that an alternative scheme is a material consideration to a decision on an application for planning permission even where, on the facts before the decision maker, there is no likelihood or real possibility of that alternative scheme eventuating. Having developed his argument thus far, he submitted that Ouseley and Moses JJ had been wrong to hold that Mount Cook's proposals were irrelevant to, as distinct from having no weight in, the council's determination of the Redevco application. He maintained that Moses J had erred in law, in stating that the only obligation on the council was to consider whether the application "left the character or appearance of the conservation area unharmed". Both propositions, he submitted, were contrary to *Nottinghamshire County Council*, the effect of which was that alternative schemes were material considerations. The fact that they were unlikely to come about went to their weight, not their materiality, and did not of itself preclude the decision maker from refusing planning permission in order to preserve the possibility of taking planning advantage of them if they did. He added that the fact that the council could not force Redevco to implement Mount Cook's scheme did not meet the point that their responsibility was to regulate the use of land in the public interest, and that a refusal of the sought planning permission could have prompted Redevco to submit a further application more in keeping with Mount Cook's proposed alternative.

[27] Mr Timothy Corner QC, for the council, submitted that Moses J had correctly found that the council had been entitled to disregard Mount Cook's alternative proposals, despite their assertion that the proposals could be undermined by the grant of Redevco's application for operational changes to the rear elevation of the building. He said that Mount Cook's proposals were irrelevant, and that even if they had been relevant, they were bound to make no difference to the result, as both Ouseley and Moses JJ had found. On the issue of materiality, he made two submissions, namely: (i) the existence of an alternative scheme is not capable of rendering an otherwise acceptable development unacceptable, and is, on that basis alone, an immaterial consideration; and (ii) further or alternatively, on the facts of the case, Mount Cook's proposals were immaterial, in that they were too vague to amount to an alternative scheme worthy of consideration, and, in any event, had no reasonable prospect of being implemented.

[28] As to the first of those submissions, Mr Corner said that the council, in determining Redevco's application, had to have regard to the development plan and to other material considerations, and if, having had regard to those matters, they found the application acceptable, they were bound to grant permission. It could not be a material consideration that there might be a better scheme. Section 70(2) of the 1990 Act does not demand a comparison between alternative projects; it does not even require that application proposals should enhance, as

distinct from preserve, the character or appearance of a conservation area in which the subject property is situate; see *South Lakeland District Council v Secretary of State for the Environment* [1992] 1 All ER 573. (As I have said, and, as it happens, the council's officers advised the council's planning subcommittee that Redevco's application proposals would themselves enhance the conservation area.) But if enhancement is required, that would not require a comparison between alternative proposals in a case such as this. He submitted that if the council had refused permission on this ground, their decision would have been judicially reviewable as irrational, because they would have failed to have considered the application on its merits. He added that the facts that Mount Cook proposed to extend the OMI to the southern part of Market Place, including the rear elevation of the building, and had threatened not to implement that proposal if Redevco's permission stood, cannot make the possibility of its proposal a material consideration when, as a matter of law, it would not otherwise be.

[29] Mr Corner's alternative submission was that even if Mount Cook's proposals are capable, in principle, of being material, they were not so in the circumstances for the reasons given by Moses J in [13] of his judgment, that Mount Cook had no power to bring them to fruition and/or that, in any event, they were of no or negligible weight. He acknowledged case law, including *South Buckinghamshire District Council*, indicating the permissibility of refusal of planning permission in order to preserve an existing use of land, or to preserve it for some desirable use. But he maintained that, unlike the circumstances of this case, there must at least be a realistic possibility of such use eventuating if permission were refused. He submitted that Mr Lockhart-Mummery's decision in *Nottinghamshire County Council* is not authority to the contrary, since there the issue was whether the inspector had been entitled to refuse permission for residential development of land that was suitable for the purpose of a new primary school, which was likely to be needed and for which it had been allocated in the local plan. He added that, even though the inspector in that case had not concluded, on a balance of probabilities, that the land would be needed for that purpose, the deputy judge had found that he had been entitled to refuse permission because, given the scarcity of land and possible need for this site for educational purposes, the policy plan's preservation of the land for that purpose was, in the circumstances, a material consideration.

*Conclusions*

[30] Mr Corner, in the course of his submission, put forward the following general propositions, which, with some slight additions, I accept as correct statements of the law, and as a useful reminder and framework when considering issues such as this. They are:

(1) in the context of planning control, a person may do what he wants with his land, provided that his use of it is acceptable in planning terms;

(2) there may be a number of alternative uses from which he could choose, each of which would be acceptable in planning terms;

(3) whether any proposed use is acceptable in planning terms depends A
upon whether it would cause planning harm, judged according to relevant planning policies, where there are any;

(4) in the absence of conflict with planning policy, and/or other planning harm, the relative advantages of alternative uses on the application site, or of the same use on alternative sites, are normally irrelevant in planning
terms; B

(5) where, as Mr Corner submitted is the case here, an application proposal does not conflict with policy, otherwise involves no planning harm and, as it happens, includes some enhancement, any alternative proposals would normally be irrelevant; and

(6) even, in exceptional circumstances, where alternative proposals
might be relevant, inchoate or vague schemes and/or those that are C
unlikely or have no real possibility of coming about would not be relevant or, if they were, should be given little or no weight.

[31] Turning to the circumstances of this case, it is clear that Redevco's application, if considered on its own, would not be harmful in a planning sense and would enhance the building and the conservation area of
which it is part. Stopping there, and still considering the application on D
its own, is more than Redevco needs to establish to justify the grant of permission. It is enough for it to show that its proposed development would not adversely affect the character or appearance of the area and is otherwise unobjectionable on planning grounds. It is not necessary for it to show that its proposals would constitute an enhancement in planning
terms: see *South Lakeland per* Lord Bridge (with whom the other law lords E
agreed), at p578c-e. However, the issue raised by Mount Cook, which was not raised in *South Lakeland*, is that the proposals in Redevco's application, if considered alongside its, Mount Cook's, alternative proposals, would, or could, be harmful in a wider planning sense of frustrating or endangering a more favourable solution for the building and the area. Put more
shortly, its case is that Redevco's proposed enhancement should yield to F
its proposed better enhancement.

[32] In my view, where application proposals, if permitted and given effect to, would amount to a preservation or enhancement in planning terms, only in exceptional circumstances would it be relevant for a decision maker to consider alternative proposals, not themselves the subject of a planning
application under consideration at the same time: for example, in multiple G
change of use applications for retail superstores called in by the Secretary of State for joint public inquiry and report. And even in an exceptional case, for such alternative proposals to be a candidate for consideration as a material consideration, there must be at least a likelihood, or real possibility, of them eventuating in the foreseeable future if the application
were to be refused. I say "likelihood" or "real possibility", because the H
words tend to be used interchangeably in some of the authorities: see, for example, *New Forest District Council* v *Secretary of State for the Environment* (1996) 71 P&CR 189 *per* Mr Nigel Macleod, sitting as a deputy High Court judge. If it were merely a matter of a bare possibility, planning authorities and decision makers would constantly have to look over their shoulders

before granting any planning application against the possibility of some alternative planning outcome, however ill-defined and however unlikely of achievement. Otherwise, they would be open to challenge by way of judicial review for failing to have regard to a material consideration, or of not giving it sufficient weight, however remote.

[33] When approaching the matter as one of likelihood or real possibility, as I have already indicated, it may often be difficult to distinguish between the concepts of materiality and weight, and both, particularly weight, are essentially matters of planning judgment. But I do not consider that a court, when considering the rationality in a judicial review sense of a planning decision, should be shy, in an appropriate case, of concluding that it would have been irrational for a decision maker to have had regard to an alternative proposal as a material consideration, or that, even if possibly he should have done so, to have given it any, or any sufficient, weight so as to defeat the application proposal.

[34] In so concluding, I have been assisted by the judgment of Laws LJ in this court in *R (on the application of Scott)* v *North Warwickshire Borough Council* [2001] EWCA Civ 315[1]. In that case, a planning committee considering an application for planning permission to construct dwellinghouses declined to consider an alternative site referred to by objectors, and granted permission. Laws LJ, with whom Aldous LJ and Blackburne J agreed, upheld the committee's decision, holding that they had been entitled to disregard the alternative site. In so holding, he stated, as a general proposition, and after reference to authorities (including the judgment of Simon Brown J, as he then was, in *Trusthouse Forte Hotels Ltd* v *Secretary of State for the Environment* (1986) 53 P&CR 293, at p299), that consideration of alternative sites would be relevant to a planning application only in exceptional circumstances. However, in his explanation of the proposition, it is plain that, in his use of the word "relevance", he had in mind both materiality and weight, a practical approach with which, as I have just indicated, I respectfully agree. Although that case concerned a possible alternative site for the sought development, and Mount Cook's proposals include alternative options for the application site, the approach of Laws LJ seems to be equally applicable. This is how he put it, in paras 30-33 of his judgment:

> 30. ... consideration of alternative sites would be relevant to a planning application only in exceptional circumstances. Generally speaking and I lay down no fixed rule... such circumstances will particularly arise where the proposed development, although desirable in itself, involves, on the site proposed, such conspicuous adverse effects that the possibility of an alternative site lacking such drawbacks necessarily itself becomes, in the mind of a reasonable local authority, a relevant planning consideration.
>
> 31. ... But even if the potentially available site... were not a *necessary* planning consideration, might a reasonable planning authority nevertheless have regarded it a *possible* planning consideration, and so should have taken it into account?

---

[1] [2001] 2 PLR 59

> 32. … I do not think so. There were no clear planning objections here (to use Simon Brown J's language [in the *Trusthouse Forte Hotels* case]). In context, the learned judge in that case, by those words, was, I think, referring to substantial objections, which were on the facts, made out. Here, there were, of course, objections, otherwise plainly we would not be here at all. There had been objections in relation to… [the alternative site] as well. If the council, as the judge held, were obliged to consider whether to have regard to the alternative site, that can only have been upon the basis that that factor might have prevailed so as to persuade the council to refuse planning permission. But, in my judgment, it was simply not capable of amounting to a good enough reason for refusing planning permission on the site in question here. Objections put forward against a planning applications such as this are, of course, judged upon their merits. If they outweigh the planning benefits of the development applied for, the application will be refused. To introduce into that equation a consideration of a different character, namely whether there would be less disbenefits on another site, could only be justified for some special reason, such, as I have said, as the existence of particularly pressing need for the development…
>
> 33. It follows, in my judgment, that no reasonable council could have treated the… [alternative site] as relevant. In those circumstances, it matters not that the council thought that they were obliged not to consider it. If they had considered it, they would have been bound to reject it.

[35] In addition, there is nothing in *South Buckinghamshire District Council* or *Nottinghamshire County Council* to support the view that where, on the facts before an inspector, there is no likelihood or real possibility of an alternative proposal proceeding if the planning application under consideration were to be refused, that it should be refused anyway, against the bare possibility of that or some other alternative, more beneficial, scheme eventuating. Indeed, such a suggestion is directly contrary to the concluding words in [36] of Mr Lockhart-Mummery's judgment in *Nottinghamshire County Council* (see [22] above), more aptly going to weight, rather than to whether the alternative is a material consideration:

> In striking the balance, the likelihood of use A actually coming about is doubtless a highly material consideration.

It would be highly harmful to the efficient and otherwise beneficial working of our system of planning control if decision makers were required to consider possible alternatives, of which, on the facts before them, there is no likelihood or real possibility of occurrence in the foreseeable future.

[36] Accordingly, I agree with the approach of Ouseley and Moses JJ and the submissions of Mr Corner. In the circumstances of this case, the alternative proposals of Mount Cook, such as they were, were not material considerations within sections 54A or 70(2) of the 1990 Act, or, if they were, they were of such negligible weight that the court was entitled to refuse permission to apply for judicial review, because the council could not reasonably have taken any notice of them.

[37] As Mr Corner put it, there is no conceivable basis upon which Mount Cook's proposals could have caused the council to reach a different decision on Redevco's application: see, for example, *Bolton Metropolitan Borough Council* v *Secretary of State for the Environment* (1990) 61 P&CR 343,

at p353 *per* Glidewell LJ, and *North Warwickshire per* Laws LJ at p65. On the contrary, if the council had refused these relatively minor alterations, their decision would have been judicially reviewable for failure to consider the application properly on its own merits.

[38] The council had had an obligation to consider Redevco's application on its own merits, having regard to national and local planning policies and any other material considerations, and to grant it, unless they considered the proposal would cause planning harm in the light of such policies and/or considerations. On the following information before the council, I do not consider that it is usurpation by the court of the council's responsibility for the planning decision, as suggested by Mr Steel, to refuse Mount Cook's claim for judicial review. That is so, even if the matter is one of weight as distinct from the materiality of the consideration.

[39] First, Mount Cook's proposals included works to the building that were different from those proposed by Redevco in its application, and Mount Cook could not implement them without Redevco's consent. Redevco had not given any such consent, and, in persisting with its application for planning permission, was clearly not minded to do so. Second, there was, and is, no evidence before the council of any prospect of Redevco giving such consent. And, contrary to Mr Steel's suggestion, I do not consider that the council had a duty to test Redevco's attitude by refusing its application in order to see whether, as a result of negotiation between the parties or otherwise, that would produce a change of heart. Third, as Mr Corner emphasised, Mount Cook's alternative proposals were "extremely inchoate". They did not take the form of a planning application, but merely, as Mount Cook's advisers described them, of "further options in the form of urban design studies for improvements". As the council's planning officer had observed, in correspondence with Mount Cook's advisers, its proposals for general improvement of the southern part of Market Place were vague, in particular "very sketchy" with regard to its proposed works to the highway, and without details for implications for traffic movement and servicing. In the circumstances, and, as I have said, Mount Cook's threat, that it would not continue with its wider proposals if Redevco were permitted to proceed with its alterations to the building, could not have a life of its own as a material consideration. That is, not only because of the lack of any likelihood or real possibility of Mount Cook being able to bring about its proposals for the building but also because there was no evidence before the council that refusal of Redevco's application would assist it in doing so.

*Issue 2: Whether the permission of external operational development, coupled with proposals to make internal alterations not in themselves susceptible to planning control, might prejudice future planning control of non-conforming change of use*

[40] This issue concerns the relationship between external operational development, for which Redevco sought and obtained planning permission, in particular, the installation of new doors and a segregated entrance lobby on the Market Place elevation, and the removal of black film from the inside of the windows on the first and second floors, and

the internal alterations that it proposed, which did not require planning permission, including the removal of an internal escalator between the second and third floors. Mount Cook maintained that the internal alterations could prejudice the continued retail use of the building, making reference to Redevco's pending separate application for permission for change of use of the upper floors of the building from retail to office and residential use. Mount Cook's expressed concern was that such internal alterations, in conjunction with the external development the subject of the application, could make it more difficult for the council thereafter to resist applications for such change of use, in conflict with policy SS2 in the local plan to protect retail floorspace in large stores trading on several floors. The planning officer, in his written report to the planning subcommittee, advised:

> Notwithstanding the objectors' concerns about the future use of the building, the Council has a duty to consider the current application on the basis on which it has been submitted ie for continued retail use. Any objections relating to the loss of retail accommodation on the upper floors would be assessed as part of a separate application.

He confirmed this advice orally to the subcommittee on their consideration of the application, observing that the application proposal would enhance the appearance of the building and the character and appearance of the conservation area, and that it had to be considered on its own merits.

[41] As Moses J observed, in [18] of his judgment, the council's planning subcommittee had clearly considered and rejected these additional arguments of Mount Cook. They had decided that the removal of film to the windows on the upper floors of the building would not prejudice the further use of those floors for retail use, and pointed out that the removal of the internal escalator was not the subject of planning control. He rejected this challenge to the permission, observing that the council had rightly decided to grant it on the basis that Redevco's proposals not only would not harm, but would also enhance, the appearance of the Market Place frontage to the building, and would not inhibit the council's planning control over any future attempts to secure non-conforming change of use. He observed that, in so far as the prospects of continued retail use of the upper floors might be lessened by the removal of an internal escalator, because such removal would not be subject to planning control, the council could not therefore prevent it. He thus considered that the council had correctly considered the application on its own merits, and that Mount Cook's expressed fears for the future were irrelevant and unarguable.

*Submissions*

[42] The main burden of Mr Steel's argument was that although Redevco could continue retail use of the upper floors of the building, and would require planning permission for any change of use, its present proposals, if they were to proceed, would undermine resistance to such change. He prefaced his detailed submissions with the observation that operational development undertaken in a building can be relevant to a determination

as to whether there has been a material change of use of the building, citing *Lake District Special Planning Board* v *Secretary of State for the Environment* [1981] JPL 363, and that internal alterations, although not themselves susceptible to planning control, might be the subject of enforcement proceedings where they are an integral part of unauthorised use of the building, citing *Somak Travel Ltd* v *Secretary of State for the Environment* [1987] JPL 630 (DC) *per* Stuart-Smith J (as he then was) at pp633-634. He claimed that such relationship between operational development and effecting or facilitating unauthorised change of use also applies, where operational development, for which permission is being considered, is "integral to an (as yet) unauthorised future use". He maintained that a decision maker, considering an application for such permission, is entitled to have regard to its practical effect upon future use, and to refuse it on the ground that it might make it hard to resist a subsequent application for a material non-conforming change of use. Applying that extension of the principle to the facts of this case, he submitted that, given Redevco's pending application for change of use of the upper floors of the building to office and residential use, the proposed inclusion in the sought operational development of new doors to the Market Place elevation and the removal of black film from the first and second floor windows would facilitate a segregated entrance lobby for the proposed change of use on the upper floors. On that basis, he maintained that the council had failed to have regard to a consideration material to their decision, and that Moses J had been wrong to dismiss the complaint on the ground that it was irrelevant.

[43] Mr Corner submitted that the judge had rightly held that the council were entitled to consider Redevco's application on its own merits and to disregard the suggestion that to grant permission on that basis would be likely to facilitate future non-conforming loss of retail use on the upper floors of the building. He submitted that such a concern was not relevant to the present application, which was not for change of use, but to the quite separate application that Redevco had pending for such change of use, and that Moses J had been correct to reject the submission as irrelevant and unarguable for the reasons he gave. He added that there is nothing in the grant of planning permission for the external works that would prevent the council from either refusing planning permission for change of use or from taking enforcement proceedings to prevent any such unauthorised use.

*Conclusion*

[44] This complaint of Mount Cook – one of "planning creep" – is barely distinguishable from its first challenge, namely that it was a material consideration for the council that the grant of permission to Redevco would, or might, frustrate an alternative scheme of enhancement – an argument to which, save in exceptional circumstances, the *South Lakeland* authority is a bar. Here, too, the correct approach for the council's planning subcommittee was the one that they adopted, namely to consider whether Redevco's proposal would cause planning harm by assessing it against doing nothing, not as against any potential enhancement that might emerge from Mount Cook's proposals. As it happens, the council's

planning subcommittee did consider and reject Mount Cook's contentions as to possible planning creep. As to Mr Steel's reliance upon *Lake District* and *Somak*, both, unlike this case, were concerned with a present non-conforming change of use, in the former as to whether internal alterations could properly be taken into account in enforcement proceedings in respect of unauthorised change, and, in the latter, whether, as an integral part of unauthorised use, they could be the subject of enforcement proceedings. And as to Redevco's pending application for change of use to mixed office and residential use on the upper floors, that is an application of which the council's planning subcommittee, considering this application for operational development, were aware, and which they can decide when the matter comes before them.

*Issue 3: Discretionary refusal of relief*

[45] If I am right in my conclusion that Mount Cook's claim is unarguable on the law and facts, the question of refusal or relief in the exercise of the court's discretion does not arise. However, in view of the conflicting submissions of counsel on this issue, I should make it plain that if it had been necessary to consider the point, I would not have refused relief in the exercise of my discretion in reliance upon the motive of Mount Cook in seeking it, namely to put pressure on Redevco to sell its lease to Mount Cook rather than, or in addition to, a genuine concern about future loss of retail use in the upper parts of the building.

[46] The essential question for a decision maker in planning matters is whether representations one way or the other, whatever the motives of those advancing them, are valid in planning terms. A collateral motive may have relevance to the reasonableness of a landlord's refusal to consent to alterations, as Mr Paul Morgan held in his judgment in the leasehold dispute between the parties that I have mentioned in [5]. But judicial review applications by would-be developers, or objectors to development in planning cases, are, by their very nature, driven primarily by commercial or private motive, rather than by a high-minded concern for the public weal. I do not say that considerations of a claimant's motive in claiming judicial review could never be relevant to a court's decision on whether to refuse relief in its discretion, for example, where the pursuance of the motive in question goes so far beyond the advancement of a collateral purpose as to amount to an abuse of process. The court should, at the very least, be slow to have recourse to that species of conduct as a basis for discretionary refusal of relief. In any event, it would, as Mr Steel pointed out, be exceptional for a court to exercise discretion not to quash a decision that it found to be *ultra vires*: see *Berkeley* v *Secretary of State for the Environment, Transport and the Regions (No 1)* [2001] 2 AC 603[1] *per* Lord Hoffmann at p616D-G, approving an observation of Glidewell LJ in *Bolton* at p343.

[1] [2000] 3 PLR 111

*Issue 4: Costs at the permission stage*

[47] The fourth issue raises a matter of considerable public importance, namely as to the guidance to be given by this court concerning the award of costs at the permission stage of claims for judicial review. The issue affects not only claimants and defendants but also interested parties and the court itself in the access that it provides to justice, having regard to the overriding objective of dealing with cases justly in CPR 1.1 and good public administration. More precisely, on the facts of this case, the issue is whether Moses J had been entitled, in the exercise of his discretion, to order Mount Cook to pay the council's costs of filing an acknowledgment of service and of successfully resisting its oral application.

[48] The issue arises under the relatively new procedure for the grant of permission for claiming judicial review, introduced by CPR 54 on 2 October 2000, supplemented by a judicial review practice direction. Both followed a review of the Crown Office, under the chairmanship of Sir Jeffery Bowman, who submitted his report (the Bowman Report) to the Lord Chancellor in March 2000. This procedure replaced the practice under RSC Ord 53 of an *ex parte* application for leave to move for judicial review, normally made on paper, but which could also be made orally at an *ex parte* hearing. A respondent, if notified of the application (*ex parte* on notice), could make representations on paper and/or, if he chose to attend and was allowed by the court to participate in a permission hearing, orally. If a respondent successfully resisted the grant of permission at an oral hearing, the court had the power to award him costs against the applicant, but it was sparing in its exercise of it. Given that practice, renewed oral applications for permission were normally heard *ex parte* and were, in any event, short. Applicants, on the whole, were able to seek relief without fear, if permission were refused, of being saddled with the respondent's costs at that stage.

[49] The new procedure involves the proposed defendant and any interested party right from the start, and is generally dealt with in the first instance as a paper application. By CPR 54.7, the claimant must serve a claim form on the defendant and any interested party within seven days of issue. By CPR 54.8, any such person "who wishes to take part in the judicial review" is required to file an acknowledgment of service. If he files an acknowledgment of service and intends, in taking part in the judicial review, to contest the claim, CPR 54.8(4) requires him to plead it in the acknowledgment of service, and to summarise his grounds for doing so.

[50] However, CPR 54 says nothing direct about the costs of filing such a document, nor, indeed, about the costs of, and incurred by, a defendant who chooses, in accordance with his entitlement under para 8.5 of the practice direction, to attend and argue his case at an oral renewal hearing. There is an indirect reference to costs in CPR 54.9. By 54.9(1), a failure to comply with the requirements as to acknowledgment of service by a party who subsequently seeks to take part in a permission hearing may, but will not necessarily, result in the court not allowing him to do so. But if he is allowed to take part, by 54.9(2), the court may take

his failure into account "when deciding what order to make about costs", a A
provision that may have, as one of its premises, that a successful defendant at the permission stage who *has* complied with CPR 54.8 should normally be entitled to his costs of filing the acknowledgment of service. Another premise may be that a defendant who has not complied with CPR 54.8, and who has not attended a permission hearing, but who later succeeds on the substantive hearing of the claim, should have some or all of his costs B
disallowed, because of his failure to comply with rule and thus to put his case to the court at the permission stage.

[51] However, regardless of the question of costs, there is now a positive obligation on a defendant or other interested party served with the claim form to acknowledge service and to consider, in doing so: (i) whether to contest the claim, and, if so, upon what grounds and at what stage; and C
(ii) if he decides to contest it, to summarise his case at the permission stage. The clear purpose of these changes, as recommended in the Bowman Report, is to enable the judge who is dealing with the application on paper to identify, at that early stage, the strengths and weakness of the proposed claim and to encourage, where appropriate, settlement of meritorious claims at the permission stage. D

[52] The only direct provision as to costs at the permission stage is that in para 8.6, when read with para 8.5, of the practice direction. Those paragraphs read:

> **8.5** Neither the defendant nor any interested party need attend a hearing on the question of permission unless the court directs otherwise.
>
> **8.6** Where the defendant or any party does attend a hearing, the court will E
> not generally make an order for costs against the claimant.

There are two important points to make about these provisions. First, on one view, when read together, they provide that, in general, a claimant should not have to pay a defendant's or other interested party's costs of attendance at a permission hearing, but they say nothing F
about the costs of filing an acknowledgment of service by a defendant who intends "to take part" in the judicial review, whether or not he subsequently attends such a hearing. On another view, they do not, or should not, distinguish between the costs of obligatory acknowledgment of service and of optional attendance at a permission hearing, whether or not the party who has filed an acknowledgment of service attends G
such a hearing: see Robert McCracken and Gregory Jones *Leach and Permission Costs* [2002] JR 4. Second, the guidance, in my view, applies to the costs of preparation for, as well as of attendance at, a hearing.

[53] The council, which intended to contest the sought judicial review, filed an acknowledgement of service, as required by CPR 54.8(2) H
and (4), and, as is plain from the facts giving rise to this appeal, attended and successfully resisted the grant of permission at the permission hearing.

[54] Before I continue, I should mention the ruling of Collins J in *R (on the application of Leach)* v *Commissioner for Local Administration* [2001]

EWHC 445 (Admin)[1], a case concerning a defendant's claim to the costs of filing an acknowledgement of service, and the effect attributed to it in the notes in the current *White Book*, at paras 54.12.5-6. The case did not concern the costs of an oral permission hearing, for there was none. The claim form and application for permission, to which the defendant had filed an acknowledgment of service, indicating his intention to contest the claim and the reasons why, had been refused by a different judge on paper. The hearing before Collins J was to consider the defendant's application for an order that the claimant should pay his costs of filing the acknowledgment of service. Although only the defendant appeared and was represented by counsel at the hearing, the claimant's solicitor wrote to the court, objecting to the proposed order, a letter that Collins J took into account. He made the order sought, expressing the view that, since the new procedure imposes on a defendant who seeks to take part in and contest a judicial review claim an obligation to file an acknowledgement of service containing a pleading of his case, it is only fair that he should be awarded his costs if, as a result, he successfully resists it at the permission stage. It is not apparent from the judgment whether, in reaching that view, Collins J's attention was drawn to paras 8.5 and 8.6 of the practice direction. This is how Collins J put it in [14] to [18] and [21] of his judgment, which was *extempore*:

> [14] The purpose of… [CPR 54.9(2)] would appear to be that where points that showed that the claim lacked merit, were not made at the permission stage, but were raised on the hearing, the court might take the view that it was not fair that the applicant should pay the extra costs that could have been avoided if only the points had been made at the earlier stage. But that, of course, only underlines the point made by Mr Corner [counsel for the defendant], that if that is one of the purposes behind the new provisions, and the requirement is there, then why should the successful party, in this case the defendant, have to bear the costs of putting forward his objections to the claim, if those objections then serve to defeat the claim? Why should he be required by the rules to incur costs which he can never recover, even if he is successful as a result of what he has done? That, submits Mr Corner, is manifestly unfair, and I agree with him. On the face of it, and having regard to the new rules, it is difficult to see that there is any sensible answer to the submission that Mr Corner has made. It seems to me that, in principle,… if a defendant incurs costs in submitting an acknowledgement of service, as required by the rules, then he ought to be able, if he succeeds, to recover his costs of so doing.
>
> [15] How much, in principle, should he be able to recover? It seems to me that it should be limited to the costs incurred in actually producing the acknowledgement, and those will obviously depend on the circumstances…
>
> …
>
> [17] But it seems to me that if this is to prevail, and if I am right in my conclusion that, in principle, costs should be awarded, it is thoroughly undesirable that there should be a need for an application, such as had to be made in this case, to obtain such costs. That, of course, only adds to any amount that is payable.

[1] [2001] 4 PLR 28

> [18] It is obvious that the rules committee is going to have to consider, in detail, the implication of this decision, but, as it seems to me, it ought to be dealt with by the judge when he deals with the permission application, and that can only happen if the application for costs is made in the body of the acknowledgement, and an indication is given as to the amount of costs which are being requested. That, of course, has to be served on the other side, who would have to have an opportunity to deal with it. A
>
> ... B
>
> [21] I am conscious... that I have not been able, since this is an *extempore* judgment and it would be equally undesirable to reserve to incur yet further costs, to have spelt out precisely what should be done for the future. One thing that seems to me to be essential is that this decision of mine, that, in principle, costs ought to be awarded, must be given wide publicity, because I suspect that claimants at the moment are simply unaware that they run the risk of orders such as this as a result of the change in the rules. C

[55] I have set out Collins J's reasoning at some length, to demonstrate that his ruling was confined to the award to a successful defendant at the permission stage of his costs of filing an acknowledgment of service. It did not extend to an award to a defendant of any of his other costs in successfully resisting a claim at the permission stage, in particular to any costs of, and/or occasioned by, an attendance at a permission hearing. D However, the editors of the *White Book*, in the following note on the decision, in the autumn 2002 and 2003 editions, at paras 54.12.5-6, appear to have given it a wider effect, suggesting that it gives a defendant a right in principle to the recovery of other costs incurred in successfully resisting the claim at the permission stage, including his costs of attendance at an oral hearing. E

> The High Court has now held that, as a result of the changes in the rules governing judicial review claims, a defendant who resists the grant of permission should, in principle be entitled to recover his costs. The High Court applied that principle to the costs incurred in filing an acknowledgment of service, and did not limit the defendant simply to recovering the costs of attendance at an oral hearing. F Defendants who wish to claim such costs should normally make the application for costs in the body of the acknowledgment of service and provide details of the amount claimed...

[56] That is plainly not what Collins J said. Nor would it flow from his reasoning that a successful defendant should be entitled to his costs G of filing an acknowledgment because he is now obliged to serve one if he wishes to take part in, and contest, a claim for judicial review. Whether or not he serves an acknowledgment of service, he has no obligation to attend a permission hearing.

[57] Moses J held that there were exceptional reasons for ordering Mount Cook to pay the council their costs of, and incurred by, the oral H renewal hearing. This is how he put it, at [82] to [85] of his judgment:

> [82] It is plain now that the court will from time to time award costs to a defendant, not only of the oral hearing but also of the acknowledgement of service, despite paragraph 8.6 of the Judicial Review Practice Direction. The notes in the White Book under 54.12.6 make this plain, as does the decision in

> A *R* v *Leach and the Commissioner for Local Administration*, a transcript of which I do not have before me, although I have seen it in the past. It does seem to me that a defendant who persists in renewing an application in circumstances such as these, where it is a highly sophisticated claimant with access to the highest possible quality legal team, pursues a claim in the face of trenchant dismissal by an experienced planning judge, forcing a local authority, funded by the local council taxpayers, to attend a full hearing, should, at the very B least, pay the costs in full of the oral hearing.
>
> [83] The more difficult question is as to whether it should pay the costs leading up to that hearing, in particular the preparation of papers and of the acknowledgment of service. *Generally, as it seems to me, there should be special features, which are not possible or indeed desirable to identify, before all the costs are borne*, merely because the rules require an acknowledgment of service C to be filed [sic]. The whole process of applying for permission for judicial review was not intended to be like ordinary litigation: the issue of a claim with issue of a defence. The mere fact that the defendant is now required to participate does not seem to me that normally where the defendant is successful he should have his costs of that acknowledgment of service and general preparation. [my emphasis]
>
> [84] There are, however, in my view, special features in this case. It is D plain that Ouseley J, a highly experienced planning judge, thought there was absolutely nothing in this case; nor do I. Although it has been skilfully argued with great attraction by Mr Steel QC, underlying it was, in my view, an absolutely hopeless attack upon the Council.
>
> [85] In those circumstances… it is right that that should be reflected by the Council having all its costs of resisting the claim…

E [58] In summary, Moses J had disagreed with what he understood from the note in the White Book was the effect of Collins J's judgment, and seemingly took the view that the guidance in para 8.6 of the practice direction, that costs should not generally be awarded to a defendant who attends a permission hearing, applied, not only to all costs of and occasioned by it but also to the costs of filing an acknowledgment of F service. On that approach, the judge had considered whether, and had held that there were, exceptional circumstances, special features for not following that guidance in relation to all the council's costs leading to their success at the permission stage. He identified three such circumstances: (i) the robustness of Ouseley J, an experienced planning judge, in his expression of refusal on the papers; (ii) the fact that Mount Cook had significant resources to mount the application and meet the consequences G of its refusal; and (iii) that it had used those resources effectively to secure a full hearing of the claim at the application stage.

*Submissions*

[59] Mr Steel asked the court to set aside the judge's order for costs, bearing in mind, in particular, that this court has now granted permission H to claim judicial review and has thereby acknowledged that Mount Cook had an arguable case. He submitted that: (i) *Leach*, whether applicable only to the costs of filing an acknowledgment of service, or of wider application, conflicts with para 8.6 of the practice direction and should be disregarded; and (ii) the circumstances relied upon by Moses J, as grounds for not following the general guidance in that paragraph, were neither

individually nor collectively sufficiently exceptional for the purpose. He submitted that the judge's order offended the overriding objective in CPR 1.1 to deal with cases justly, in that the permission stage in judicial review proceedings is intended to provide claimants with, among other things, a quick and relatively cheap mechanism for initiating and testing the arguability of their challenges to the decisions of public authorities. A practice of awarding costs against unsuccessful claimants, save in truly exceptional cases, would, he said, provide a disincentive to potential claimants that would run counter to that objective.

[60] In making that submission, Mr Steel also drew on the intention, expressed in paras 18-25 of the Bowman Report, to maintain the permission stage and to introduce defendants' acknowledgements of service in order to facilitate early and more efficient weeding out of unmeritorious applications, and consideration by defendants of the legality of their conduct called into question by the claim. He drew particular attention to the indication in para 24 of the report that, in recommending the introduction of acknowledgements of service, it was not expected that defendants or other interested parties should incur substantial expense at the permission stage. And it was significant, he said, that the report had contained no recommendation for the grant of costs against unsuccessful claimants at that stage. In short, he submitted that the intention of the Bowman Report recommendations was to produce a "quick fix" one way or another, not to encourage full-scale "preliminary hearings", conducted as if they were hearings of substantive claims for judicial review.

[61] Mr Steel distinguished the apparently contrary ruling of Collins J in *Leach* as limited to its own facts, namely as to the costs of filing an acknowledgement of service. But he maintained that it should also not be given general effect, either as to the costs of filing an acknowledgment of service or in the wider sense seemingly given to it in the note in the *White Book*. To do so, he said, would run counter to para 8.6 of the practice direction and the thinking in the Bowman Report giving rise to it.

[62] As to the first of Moses J's reasons for ordering Mount Cook to pay the council's costs, namely Mount Cook's renewal of the application for permission after Ouseley J's robust refusal, Mr Steel pointed out that CPR 54.12(3) gives a claimant a right to seek reconsideration of his application by way of an oral hearing. He submitted that, in the absence of guidance in CPR 54, a claimant should not be penalised in costs for exercising a right expressly provided for by it, and in respect of which the practice direction provides that, where a defendant attends such a hearing, a claimant (implicitly, whatever the outcome) will not generally be penalised in costs. He argued that neither the robustness of the refusal on paper nor the relevant experience of the judge is exceptional; paper refusals are often robust and the Administrative Court judges responsible for them are all specialists and/or experienced in this field. Second, as to Moses J's reliance upon the resources of Mount Cook, Mr Steel submitted that he had been wrong in law and as a matter of public policy to rely upon such a factor because it put claimants with means at a disadvantage to claimants of lesser means when seeking relief from public wrongs.

[63] Mr Corner's response to these arguments was that a judge's discretion as to costs in this, as in most, proceedings is wide, notwithstanding paras 8.5 and 8.6 of the practice direction. He drew attention to the wide discretion in section 51 of the 1981 Act, to which there is no relevant contrary provision in this context apart from the practice direction, and to CPR 44.3(2), which provides that, generally, payment of costs follows the outcome.

[64] As to the costs of preparing and filing an acknowledgment of service and of the preparation for attendance at any oral renewal hearing, Mr Corner submitted that para 8.6 of the practice direction does not apply to the former, and there is therefore no bar to their recovery, whatever the true effect of that guidance as to other costs of, and occasioned by, an oral renewal hearing. He supported the generality of the note in the *White Book* as to the effect of Collins J's judgment in *Leach,* and disagreed with Moses J's view, in [83] of his judgment (see [57] above), that a successful defendant should not generally have such costs. He maintained that it followed from *Leach* that a defendant who has filed an acknowledgment of service and has indicated, pursuant to CPR 54.8(4), his intention to contest the claim, with a summary of his grounds for doing so, and who has succeeded on that "pleading" at the permission stage, should have his costs. He submitted, in relation to these and to the costs of attendance at such a hearing, that, as a general principle, it is unfair that a defendant should have to bear the costs of putting forward his objections, in whatever form, to an unarguable claim; it is simply a waste of public funds.

[65] In relation to the costs of filing an acknowledgment of service, Mr Corner drew important support from the judicial review pre-action protocol, adopted with effect from 4 March 2002. It provides, save in urgent cases, for a letter before claim and for the proposed defendant to respond setting out his stance, and, if contrary to the proposed claim, summarising the reasons for it. Failure by the proposed defendant to respond could be met with a sanction in costs. He submitted that an intending claimant could, therefore, follow this procedure without fear of incurring liability for a defendant's costs and, if a defendant provided no adequate response, without fear of being ordered to pay the costs of a defendant's acknowledgment of service. He pointed out that Mount Cook and the council had followed the protocol and submitted, therefore, that there was no reason why the council should have to bear the costs of the wholly avoidable need of preparation of an acknowledgment of service.

[66] However, in respect of all the council's costs, including those of filing the acknowledgment of service and of attendance at the oral renewal hearing, he adopted the judge's reasoning that, in any event, there had been exceptional circumstances, or "special features" as the judge called them, justifying an exercise of his discretion to award them all against Mount Cook. On this partly alternative submission, he relied upon the following circumstances: (i) Mount Cook's claim was hopeless; (ii) given the calibre of its legal representation, it must have known beforehand that it was hopeless; (iii) it had effectively had the benefit of a full, substantive, hearing of the claim; (iv) it was well able to bear the costs; and (v) in the

circumstances, it was right that the council, as a public body, should have all their costs of resisting the claim.

*Conclusions*

[67] The starting point, it seems to me, is the general provision in section 51 of the 1981 Act that, subject to any contrary statutory enactment or rules of court, costs are in the discretion of the court. There is no statutory provision or rule of court, in particular in CPR 54, removing that discretion. The nearest to intrusions on it are the general rule in CPR 44.3(2), that costs, where awarded, should generally follow the event and, in this context, paras 8.5 and 8.6 of the practice direction. The CPR are made by the Civil Procedure Rule Committee, and are made by statutory instrument, pursuant to sections 1 and 2 of the Civil Procedure Act 1997. Practice directions, in general, supplement the CPR, and are made by the head of the appropriate division of the High Court under his or her inherent jurisdiction. They are recognised by the 1997 Act, and, for example, in paras 3 and 6 of section 5(1) of Schedule 1 may, in certain circumstances, have the effect of provisions that could otherwise be made by way of CPR: see also CPR 8.1(6)(b). Such circumstances do not apply to paras 8.5 and 8.6 of the judicial review practice direction. I use the word "recognised" deliberately, for I doubt whether it is correct to assert, as a generality, as do the authors, in para 1.3.2 of the May 2000 edition of the *Queen's Bench Guide*, that practice directions are made "pursuant" to statute, or that they have the same authority as the CPR. As the guide itself asserts, in the case of any conflict between the two, the CPR prevail. To that already somewhat cumbrous and confusing three-tier hierarchy of rules and guidance for civil litigants – statutory, CPR and practice directions – there has now, as I have indicated, been added a fourth, in the case of judicial review, in the form of the pre-action protocol.

[68] As to practice directions, what is important is that all involved in the areas of administration of justice for which they provide, including claimants in judicial review proceedings, should be able to rely upon them as an indication of the normal practice of the courts, unless and until amended. However, they differ from the CPR in that they: (i) in general provide guidance that should be followed, but do not have binding effect; and (ii) should yield to the CPR where there is clear conflict between them.

[69] The guidance given in para 8.6 of the judicial review practice direction, that a claimant at a permission hearing will not generally be ordered to pay the costs of a defendant or any other party who attends, does not conflict with CPR 54, which, as I have said, makes no direct provision as to the award of costs at the permission stage. And it is, as Mr Steel pointed out, of a piece with the practice of the courts under the old Ord 53 of the Rules of the Supreme Court when read with section 51 of the 1981 Act: see, for example, *R* v *Honourable Society of the Middle Temple, ex parte Bullock* [1996] ELR 349 *per* Brooke J (as he then was) at p359C; and McCracken and Jones *op cit*. But the great change wrought by CPR 54 over RSC 53 is the requirement of service of the initial claim and of a response by an acknowledgment of service, and of an entitlement, not an obligation,

on the part of the proposed defendant, if he has so responded, to attend and to make representations at any oral renewal hearing.

[70] This new regime may be compared and contrasted with the provisions of CPR 52 and CPR 52 PD, which provide, in the case of applications for permission to appeal, for applicants to notify respondents of the application. However, unless the court otherwise directs, they do not oblige respondents to do anything unless and until notified that permission has been given. And, although clearly envisaging that an order for costs may be made at a permission hearing, they do not in terms seek to guide or fetter that discretion: see *Jolly* v *Jay* [2002] EWCA Civ 277, *per* Brooke LJ in [48], giving the judgment of the court. See also, and compare, the regime for seeking permission to proceed with a statutory planning appeal against the dismissal of an enforcement notice, under section 289 of the 1990 Act, in respect of which this court held, in *R* v *Secretary of State for Wales, ex parte Rozhon* (1993) 91 LGR 667, that, as a general rule, the costs of the application should follow the event.

[71] Here, the express discouragement in para 8.6 of the practice direction of the award of costs against claimants, whether successful or unsuccessful, at the permission stage is, in my view, a clear indication, in conformity with the Bowman Report recommendation, that if a defendant or other interested party chooses to attend and contest the grant of permission at a renewal hearing, the hearing should be short and not a rehearsal for, or effectively a hearing of, the substantive claim. The objects of the obligation on a defendant to file an acknowledgment of service setting out, where appropriate, his case are to: (i) assist claimants with a speedy and relatively inexpensive determination by the court of the arguability of their claims; and (ii) prompt defendant public authorities to give early consideration to and, where appropriate, to fulfil their public duties. It would frustrate those objects to discourage would-be claimants from seeking justice by the fear of a penalty in costs if they do not get beyond the permission stage, or to clog up that stage with full-scale rehearsals of what would be the substantive hearing of a claim if permission were granted. Thus, not only the statutory scheme, as supplemented by the practice direction and the pre-action protocol, but also the public law context is different from that governing the generality of civil law proceedings, differences that suggest the need for, and intention to provide, a different costs regime in such cases.

[72] Accordingly, I see no good reason, in law or practice, why the guidance given in para 8.6 of the practice direction should not be followed in this and all cases in which a defendant or other interested party to a judicial review claim files an acknowledgment of service and attends and successfully resists it a permission hearing. Generally, that is, save in exceptional circumstances costs of, and occasioned by, such attendance should not be awarded against a claimant.

[73] It follows that judges before whom contested permission applications are listed, and in their conduct of them, should discourage long hearings and/or the filing by both parties of voluminous documentary evidence for consideration at them. In short, they should not allow the court

to be sucked into lengthy and fully argued oral hearings that transform the process from an inquiry into arguability into that of a rehearsal for, or effectively, an expedited and full hearing of the substantive claim.

[74] But where does that general rule leave *Leach* and the costs of filing an acknowledgment of service upon which a defendant has relied and followed through by successfully resisting the claim at the permission stage? As I have said, as a result of the note in the *White Book*, the ruling of Collins J in *Leach* appears to be regarded as an authority for the proposition that a defendant who successfully resists the grant of permission should, as a matter of principle, be entitled to his costs, not only of filing an acknowledgment of service as required by CPR 54.8 but also of his preparation for, and attendance at, any permission hearing. In fact, as Mr Steel observed, there had been no permission hearing in that case. The only hearing was of an application, by an unopposed defendant, for an order that the claimant should pay his costs of filing of the acknowledgment of service. It was not, therefore, a case that would have engaged para 8.6 of the practice direction, since, when read with para 8.5, the guidance that a defendant or other interested party attending an oral permission hearing should not generally have his costs, clearly applies only to the costs of, and occasioned by, his attendance at such a hearing. Given that distinction and the absence of any such constraint on the narrower issue before him, there was, with respect, good sense in Collins J's recourse to the obligation in CPR 54.8 to file an acknowledgment as a reason for requiring a claimant to pay the costs of that initial procedural step. Different considerations, which he did not have to consider, would obviously apply to the costs of a permission hearing at which a defendant who intends "to take part in the judicial review" chooses voluntarily to attend and orally to argue his case.

[75] There are obviously some practical aspects of the *Leach* approach that, as Collins J indicated in [17] to [19] of his judgment, require urgent attention. I would suggest, in the first instance, that the Civil Procedure Rule Committee may wish to look at the matter (if it has not already begun to do so), with a view to an amendment of CPR 54.8, or to the prompting of an amendment of the practice direction, to provide appropriate machinery for claiming and the award of costs of filing an acknowledgment of service and for contesting such an award in an appropriate case.

[76] Accordingly, I would hold the following to be the proper approach to the award of costs against an unsuccessful claimant, and to the relationship of the obligation in CPR 54.8 upon a defendant "who wishes to take part in the judicial review" to file an acknowledgment of service with the general rule, in para 8.6 of the practice direction, that a successful defendant at an oral permission hearing should not generally be awarded costs against the claimant:

(1) The effect of *Leach*, certainly in a case to which the pre-action protocol applies, and where a defendant or other interested party has complied with it, is that a successful defendant or other party at the permission stage who has filed an acknowledgment of service, pursuant to

A CPR 54.8, should generally recover the costs of doing so from the claimant, whether or not he attends any permission hearing.

(2) The effect of para 8.6, when read with para 8.5, of the practice direction, in conformity with the long-established practice of the courts in judicial review and the thinking of the Bowman Report giving rise to the CPR 54 procedure, is that a defendant who attends and successfully resists B the grant of permission at a renewal hearing should not generally recover from the claimant his costs of, and occasioned by, doing so.

(3) A court, in considering an award against an unsuccessful claimant of the defendant's and/or any other interested party's costs at a permission hearing, should depart from the general guidance in the practice direction only if he considers that there are exceptional circumstances for doing so.

C (4) A court considering costs at the permission stage should be allowed a broad discretion as to whether, on the facts of the case, there are exceptional circumstances justifying the award of costs against an unsuccessful claimant.

(5) Exceptional circumstances may consist in the presence of one or more of the features in the following non-exhaustive list:

D (a) the hopelessness of the claim;

(b) the persistence in it by the claimant after having been alerted to facts and/or of the law demonstrating its hopelessness;

(c) the extent to which the court considers that the claimant, in the pursuit of his application, has sought to abuse the process of judicial review for collateral ends – a relevant consideration as to costs at the E permission stage, as well as when considering discretionary refusal of relief at the stage of substantive hearing, if there is one; and

(d) whether, as a result of the deployment of full argument and documentary evidence by both sides at the hearing of a contested application, the unsuccessful claimant has had, in effect, the advantage of an early substantive hearing of the claim.

F (6) A relevant factor for a court, when considering the exercise of its discretion on the grounds of exceptional circumstances, may be the extent to which the unsuccessful claimant has substantial resources that it has used to pursue the unfounded claim, and that are available to meet an order for costs.

(7) The Court of Appeal should be slow to interfere with the broad G discretion of the court below in its identification of factors constituting exceptional circumstances, and in the exercise of its discretion as to whether to award costs against an unsuccessful claimant.

[77] Such an approach seems to me to accord with public policy in providing ready access to the courts by individuals or bodies seeking relief from and/or to draw attention to actual or threatened transgressions of the H law by public bodies, while, in exceptional cases, protecting those bodies, and the public that funds them, from unnecessary, burdensome and costly substantive litigation. If properly and consistently applied by the courts, I can see nothing about it that would, as Mr Steel suggested, undermine the fairness and probity of judicial review as a means of control of the administration or run contrary to Article 6.1 of the European Convention

on Human Rights, Lord Woolf's civil justice reforms or the adoption of A
them in this context in the Bowman Report.

[78] It follows, in my view, that Moses J had erred in applying the same test to the council's costs of filing the acknowledgment of service as to their costs of, and occasioned by, attendance, and contesting the arguability of the claim at the permission hearing. He could, and should, have applied the discretion available to him to award such costs on the *Leach* principle B
and untrammelled by the guidance in para 8.6 of the practice direction. And, although for want of access to the report of *Leach* at the time of giving judgment, he might have been misled by the note in the White Book as to the breadth of Collins J's ruling, it is plain from his reasoning in [82] to [85] of his judgment (see [57] above) that he exercised the para 8.6 discretion in the conventional way by looking for, in his words, "special C
features" before departing from the general course of making no order in attendance costs cases. Against the argument that he might not have applied the same rigorous test – in [82] of his judgment – to the costs of the oral hearing as to those of preparation for it, in particular, the filing of the acknowledgment of service – in [83] – his reasoning was essentially the same for both. In his emphasis on the latter in [83] and [84], he was merely D
responding, albeit unnecessarily, to the approach of Collins J to the costs of filing an acknowledgment of service.

[79] It follows, in my view, that Moses J, regardless of his possible misunderstanding of the breadth of *Leach*, had identified and applied the correct test under para 8.6 to the council's costs of, and occasioned by, their attendance at the permission hearing. He had been entitled to order E
Mount Cook to pay the whole of those costs, within the range of discretion still permitted to him by that provision, and by the conventional long-standing practice of the courts to which it gave expression. What amounts to "exceptional circumstances" for not following the general rule may vary considerably according to the circumstances of the case, including the strength or weakness of the application and the respective conduct and F
circumstances of the parties. Here, the combination of the hopelessness of the application, the clear intention and ability of Mount Cook to deploy its considerable resources in attempting to use this public law route of exerting commercial pressure on Redevco to surrender its lease, and its use of those resources effectively to secure a full hearing of the claim at the application stage were clearly capable of amounting to exceptional G
circumstances for not following the general rule, and the judge had been entitled so to find.

[80] As to the hopelessness of the application, against Mr Steel's suggestion that it cannot have been that hopeless given this court's grant of permission on a renewed oral application, I should mention that: (i) in refusing permission on the papers, I was of, and expressed, the H
view that the claim was hopeless, either because of the immateriality of Mount Cook's alternative proposal or because of its lack of weight; (ii) on the oral renewal of application for permission, the court, unlike at this substantive hearing, had only heard counsel for Mount Cook; and (iii) at that hearing, both members of the court had been of the same

view, but had decided, with some hesitation, to grant permission, having regard also to the claimed uncertainty of the law as to the materiality of alternative proposals in planning applications, and as to costs at the permission stage. Having now heard both sides, I am confident that I was right first time; it was, as Ouseley J made plain in his reasons for refusal on paper, a hopeless claim.

[81] As to the considerable resources of Mount Cook and its deployment of them in initiating and persevering with the claim, I should add that, in my view, there are different considerations when a court is asked to consider the grant of costs against an unsuccessful claimant at the permission stage from those as to discretionary refusal of relief at the end of a substantive hearing: see [47] above. At the refusal of permission stage, the claimant has lost on the merits, and, as in any other case where the award of costs lies in the discretion of the court, the conduct and motive of an unsuccessful party, in having pursued unmeritorious litigation for some collateral aim, is capable of being a relevant consideration to the exercise of that discretion. As a result of the refusal of permission, there is no underlying legal justification for the claim to intrude on the manner of its exercise.

[82] As to the full hearing point, had the matter continued to a substantive hearing with the same result, Mount Cook would undoubtedly have had to pay the council's costs. There is force in Mr Corner's argument that it would have been unfair for Mount Cook to have had the benefit of fully testing its case and to have lost, without any risk of exposure to the council's costs. In addition, there is a public interest in considering the award of costs against an unsuccessful claimant in such circumstances if, as was clearly intended by the Bowman Report and the framers of paras 8.5 and 8.6 of the practice direction, lengthy contested permission hearings of this sort should be discouraged.

[83] Accordingly, I would reject all the grounds relied upon by Mount Cook, and would dismiss its appeal and its claim for judicial review.

**CLARKE LJ:**

[84] I agree.

**JONATHAN PARKER LJ:**

[85] I also agree.

*Appeal and claim for judicial review dismissed.*

# R (on the application of Gavin) *v* Haringey London Borough Council and another

[2003] EWHC 2591 (Admin)

QUEEN'S BENCH DIVISION: ADMINISTRATIVE COURT
*Richards J*
5 November 2003

**Judicial review – Relief – Discretion – Planning application – Town** A
**and Country Planning (General Development Procedure) Order 1995 – Publicity for development – Planning permission granted – Failure to comply with requirements of proposed development – Non-compliance with article 8(4) – Delay – Whether failure to comply with 1995 publicity requirements causing substantial prejudice to claimant**

B

In 1999, W Ltd acquired land for the purposes of redevelopment, and applied for planning permission in September 2000. The claimant lived opposite the development site and did not become aware of either the application or the permission being granted until April 2003. In May 2003, he issued proceedings for judicial review seeking to quash the council's decision to grant planning permission. He contended, *inter* C
*alia*, that the council had failed to comply with the publicity requirements pursuant to the Town and Country Planning (General Development Procedure) Order 1995, resulting in the local residents not being properly informed of the planning application. Article 8(4) provided for the need to publicise applications: (a)(i) by display on or near to the land relating to the application, or (ii) by serving notice on any adjoining owner or occupier; and (b) by local advertising. He also maintained that D
the council had a discretion to proceed under (i) or (ii), and that the discretion had to be exercised in accordance with the spirit and purpose of the enabling statute and, in particular, for reasons relevant to the achievement of the statutory purpose. The council had also failed to consider whether the planning application required an environmental impact assessment pursuant to the Town and Country Planning E
(Environmental Impact Assessment)(England and Wales) Regulations 1999. The council accepted that they had made a legal error, but refused relief, under section 31(6) of the Supreme Court Act 1981, on the ground of substantial hardship or prejudice to W Ltd, and detriment to good administration.

F

*Held* Declaratory relief by way of damages was awarded.

It is doubted that article 8(4)(a) imposes an obligation on the council as to which method is best used to give notice of the application to those likely to be interested. Sufficient notice is given by a combination of: (a) either of those methods; and (b) local advertisement: see [27]. In circumstances where the claimant did not receive notification of the planning application, the admitted failure to comply with article 8(4)

A would have been a sufficient reason for quashing planning permission, provided there had been a timely challenge: see [30]. The claimant was sufficiently prejudiced by the failure to comply with article 8(4)(b); he had been denied the opportunity to make representations: see [31]. The council also failed to consider whether the development was likely to have significant effects upon the environment by virtue of factors B such as its nature, size or location, pursuant to the EIA Regulations: see [34]. Had that failure to comply with the EIA Regulations been raised in a timely challenge, it would inevitably have led to the quashing of the planning permission in accordance with the principles laid down in *Berkeley* v *Secretary of State for the Environment, Transport and the Regions (No 1)* [2000] 3 PLR 111: see [35]. Substantial weight was C attached to the council's failure to comply with the publicity requirements and the EIA requirements. These were serious procedural errors, that had denied the claimant an opportunity to make representations opposing the development. The claimant was not to be blamed for the delay. There was undue delay within the meaning of section 31(6) of the 1981 Act, and to quash the planning permission would cause D very substantial hardship or prejudice to W Ltd. A declaration was granted to the claimant that the council had failed to comply with the relevant publicity and EIA requirements, but it did not affect the planning permission itself or the works carried out pursuant to it: see [91].

**Cases referred to in the judgment**

E *Amministrazione delle Finanze dello Stato* v *San Giorgio SpA* C199/82 [1983] ECR 3595

*Berkeley* v *Secretary of State for the Environment, Transport and the Regions (No 1)* [2001] 2 AC 603; [2000] 3 WLR 420; [2000] 3 All ER 897; (2001) 81 P&CR 35; [2000] 3 PLR 111; [2001] JPL 58, HL

*De Geouffre de la Pradelle* v *France* [1992] A253-B

F *FG Whitley & Sons Co Ltd* v *Secretary of State for Wales* (1992) 64 P&CR 296; [1992] 3 PLR 72; [1992] JPL 856, CA

*O'Reilly* v *Mackman* [1983] 2 AC 237; [1982] 3 WLR 1096; [1982] 3 All ER 1124, HL

*R (on the application of Burkett)* v *Hammersmith and Fulham London Borough Council; sub nom R* v *Hammersmith and Fulham London Borough Council, ex parte Burkett; Burkett, Re* [2002] UKHL 23; [2002] 1 WLR 1593; [2002] 3 All ER 97; [2002] 2 PLR 90

G *R (on the application of Lichfield Securities Ltd)* v *Lichfield District Council; sub nom R* v *Lichfield District Council, ex parte Lichfield Securities Ltd* [2001] EWCA Civ 304; [2001] 3 PLR 33; [2001] JPL 1434; [2001] PLCR 32, CA

*R* v *Dairy Produce Quota Tribunal for England and Wales, ex parte Caswell* [1990] 2 AC 738; [1990] 2 WLR 1320; [1990] 2 All ER 434, HL

*R* v *Flintshire County Council, ex parte Somerfield Stores Ltd* [1998] PLCR 336, QBD

*R* v *Newbury District Council, ex parte Chieveley Parish Council* (1998) 10 Admin H LR 676; [1998] EGCS 131, CA

*R* v *North West Leicestershire District Council, ex parte Moses (Costs)* unreported 12 April 2000

*R* v *Tower Hamlets London Borough Council, ex parte Chetnik Developments Ltd* [1988] AC 858; [1988] 2 WLR 654; [1988] 1 All ER 961; [1988] 2 EGLR 195; [1988] 26 EG 69, HL

**Judicial review** A

This was a claim for judicial review by the claimant, Seamus Gavin, in respect of a grant of planning permission by the defendants, Haringey London Borough Council, for retail development to the interested party, Wolseley Centres Ltd.

*Robert McCracken QC and Angela Ward (instructed by Gavins Solicitors) appeared for the claimant, Seamus Gavin.* B

*Geoffrey Stephenson (instructed by the solicitor to Haringey London Borough Council) represented the defendant council.*

*Peter Goatley (instructed by BPE, of Cheltenham) represented the interested party, Wolseley Centers Ltd.*

*The following judgment was delivered.* C

**RICHARDS J:**

[1] This is a claim for judicial review of the grant of planning permission by Haringey London Borough Council (the council) to Wolseley Centers Ltd (Wolseley), on 22 September 2000, for a development at 460 Archway Road, Highgate. For reasons explained below, the claim was lodged only D on 30 May 2003, over two-and-a-half years after the grant of planning permission. The council and Wolseley opposed permission on the ground of delay, but permission was granted by Elias J after a contested hearing in June 2003. Nevertheless, delay and its effects remain at the heart of the case, since both the council and Wolseley accept that the process leading to the grant of planning permission was affected by legal error, but submit that relief should be refused, pursuant to section 31(6) of the E Supreme Court Act 1981, on grounds of substantial hardship or prejudice to Wolseley and detriment to good administration.

*Factual background*

[2] The site in question runs between a railway yard and Archway Road and is 0.88ha in area. It was previously owned by London Underground F and leased out for use by a builders' merchant business and a vehicle hire and sales business. In 1999, ownership of the site passed to Wolseley, which then applied for planning permission for what is described in the grant itself as "redevelopment of the site to provide general building supplies warehouse, sales area, bag and board store, hire centre, plumbing and draining and retail bathroom equipment showroom". Wolseley's evidence G is that the redeveloped site is intended to house an existing building centre operation, together with a timber centre, hire centre, plumbing centre and drainage centre. The floorspace of the proposed buildings is 2,750m$^2$.

[3] The claimant lives (and runs a solicitor's office) at 473 Archway Road, directly opposite the southern end of the site. His evidence is that he was unaware of the application for planning permission or of the grant H of planning permission until April 2003.

[4] According to the council's records, a total of 94 written notices of the application were sent out in November 2000 to occupiers of properties in the area, including the claimant, and to other consultees. The council feel unable, however, to dispute the claimant's evidence that he did not

receive any such notice. From the claimant's researches, it would appear that some of the notices were sent to non-existent addresses and that, in many other cases, the notices were not received. It is not necessary to go into the detail of this. Suffice it to say that, in my view, an attempt was made by the council to send notices to local residents but, in practice, for whatever reason, it was not very successful, though among those who did receive notification were the Highgate Society and the Highgate Conservation Area Advisory Committee, which made representations objecting to the proposal.

[5] The planning application was also advertised in a newspaper called the *Haringey Independent*. Unfortunately, as is now accepted by the council, the circulation of that newspaper did not include the postcode area where the site is situated.

[6] There was no site notice advertising the planning application.

[7] The net result is that there was much less awareness of the planning application among local residents than there should have been, and than had been intended by the council.

[8] The planning application was considered and approved by a subcommittee of the council on 6 April 2000. The report to the subcommittee stated that there had been 85 consultees and one objection. The source of the figure of 85 consultees is unclear – in contrast, the 94 written notices to which I have already referred – but, in any event, it is likely that substantially fewer than 85 people had actually received notification of the application. I shall deal later with an issue raised concerning the subsequent dismissal of the author of the report to the subcommittee.

[9] The formal grant of planning permission was on 22 September 2000. There was subsequently some discussion about the scope of the permission granted, which led to Wolseley giving, in May 2002, a unilateral undertaking, under section 106 of the Town and Country Planning Act 1990, that the retail part of the development would not be used otherwise than for the retail sale of a specified range of goods (including products as diverse as DIY products, building materials, bedding and household textiles, audio tapes/CDs etc, and wines, spirits and confectionery).

[10] Work commenced on the site in August 2002, but only for a period of a week or so, and at a time when the claimant was abroad on holiday. During that period, Wolseley's contractor cleared some trees on the southern part of the site and demolished a building described as the former coke office. The contractor then went into liquidation and the work ceased. The claimant did not see the work being carried out and did not have it brought to his attention by any of his neighbours, although the demolition of the coke office had been noticed by Mrs Smith, a lady who lived at 497 Archway Road. The claimant's evidence is that, on his return from holiday, he noticed what looked to him like some pruning of the trees on the southern end of the site, but was not concerned about it, since he thought it was normal management of the trees by London Underground.

[11] In November 2002, hoardings were erected along about 250ft of the frontage at the northern end of the site. The claimant says that it did not give the impression of impending development works, as there were none of the usual developers' paraphernalia, such as safety notices and details of architects and contractors.

[12] In March 2003, Wolseley selected a second contractor, Brennan Ltd, to carry out the work of redevelopment. Brennan arranged for an excavator to be brought onto the site on or about 13 March, for the purpose of digging trial pits in order to determine ground conditions. Wolseley's evidence is that the trial pits were dug in the morning and were back-filled by lunchtime.

[13] Referring to what must have been the same occasion, the claimant says that he saw an articulated digger working on the site directly opposite his home, removing trees and clearing the site. He immediately contacted the council's planning department by telephone. He was told to put his complaint in writing, which he did. He received an acknowledgement, and then, on 27 March, a letter stating that no record of the grant of planning permission for works on the site had been found. This was followed by a letter from the council, dated 7 April, stating that a "valid" planning permission had, in fact, been granted to develop the site. That was the first time the claimant became aware of the planning permission or the application for it.

[14] Over the next few weeks, the claimant sought to ascertain in detail the circumstances in which planning permission had been granted. He was in contact with the council by telephone and in correspondence, and he attended the council's offices to inspect the planning file and to meet a council officer. He made enquiries about the circulation of the *Haringey Independent* in which the relevant advertisement had been placed. He carried out a survey of the addresses to which, according to the council's records, notices of the application had been sent. He made detailed enquiries of interested parties that had, apparently, received notice of the application. He raised his concerns about the council's failure to comply with relevant procedural requirements and, having failed to obtain a satisfactory response, on 14 and 19 May, he sent letters before claim.

[15] During the same period, the claimant also contacted Wolseley. On 15 April, he spoke by telephone to Mr Mason, Wolseley's development manager. According to Mr Mason, the claimant "telephoned... to say that he had good cause to question the validity of the planning permission granted by Haringey council in September 2000 and requested a copy of the plan showing the development". Mr Mason sent a copy of the plan the following day. On 30 April, the claimant wrote to Wolseley's lawyers to confirm a telephone conversation on the same day, in which "we informed you that we have a good case for challenging the grant of planning permission". The letter gave details of counsel who had been instructed and whose advice was said to be expected shortly. It stated that "you should bear in mind that if we are forced to stop this development through the courts, your client will face adverse media coverage", and concluded: "Accordingly, we give your

A client the opportunity to consider their position with a view to their stopping work immediately."

[16] Wolseley did not stop the work. I shall have to deal later with the various implications of this.

[17] The claimant lodged the claim for judicial review on 30 May 2003. He did not seek interim relief. The grant of permission by Elias J was
B on 23 June. The substantive hearing was expedited and was listed for mid-October.

[18] Following Elias J's grant of permission to apply for judicial review, Wolseley submitted a fresh application for planning permission, materially identical to the permission already granted. It appeared, at one time, that a decision on the fresh planning application might be
C reached by the planning subcommittee before the substantive hearing of the judicial review application. On that basis, the council applied for a three-month stay of the judicial review proceedings. This was rejected by Collins J on 6 October, largely because, by that time, it had become clear that a decision on the fresh planning application would not be taken until December 2003 or January 2004. The current position is that the fresh
D planning application is likely to be considered in a development control forum (a non-statutory procedure triggered by a petition of at least 25 objectors, and involving a meeting between objectors, developer and council officers) before any decision is taken on it by the relevant subcommittee. In the circumstances, it is clear that the challenge to the existing planning permission must be resolved without waiting for a
E decision by the council on the fresh application.

*Issues*

[19] Although there is only limited dispute about them, I shall deal first with the grounds of challenge to the grant of planning permission, namely: (i) failure to comply with the publicity requirements in the Town and Country Planning (General Development Procedure) Order 1995; and
F (ii) failure to comply with the requirement to consider whether the application required an environmental impact assessment (EIA) pursuant to the Town and Country Planning (Environmental Impact Assessment)(England and Wales) Regulations 1999. Mr Robert McCracken's skeleton argument referred to a third ground of complaint, namely that the terms of the grant of planning permission were vague, ambiguous and
G so uncertain as to be unlawful; but this was not part of the pleaded claim and, following an early warning shot from Mr Geoffrey Stephenson, was not pursued, in this form, in Mr McCracken's oral submissions.

[20] The relief sought by the claimant is the quashing of the planning permission or, alternatively, some form of declaratory relief that, as it was put, would not affect the validity of the planning permission but
H might provide a basis for a claim in damages and would provide the claimant with some satisfaction. No damages claim is pursued in the present proceedings.

[21] It will be necessary to consider the principles governing the exercise of the court's discretion with regard to the grant or withholding of relief in a case where there has been delay in bringing the challenge,

and then to consider the individual factors relevant to the exercise of that discretion on the facts of the present case. Such factors include: (i) the nature of the legal errors and the conduct of the council; (ii) the period of delay and the conduct of the claimant; (iii) hardship or prejudice to Wolseley if relief is granted; (iv) other issues concerning the conduct of Wolseley; and (v) detriment to good administration if relief is granted. Brief consideration must also be given to arguments raised under the Human Rights Act 1998.

[22] Finally, I shall have to balance the various factors and reach a considered judgment on how the court's discretion should be exercised.

*Failure to comply with publicity requirements*

[23] Article 8(4) of the Town and Country Planning (General Development Procedure) Order 1995[1] provides:

> In the case of an application for planning permission which is not a paragraph (2) application, if the development proposed is a major development the application shall be publicised by giving requisite notice –
>
> (a)(i) by site display in at least one place on or near the land to which the application relates for not less than 21 days, or
>
> (ii) by serving the notice on any adjoining owner or occupier, and
>
> (b) by local advertisement.

[24] Mr McCracken does not pursue, for present purposes, a contention that this was a para (2) application. The development proposed was a "major development" by reason of the proposed floorspace. Article 8(4) can, therefore, be treated as the provision with which the council had to comply by way of publicity for the application.

[25] It is common ground that there was a failure to comply with article 8(4), although there is a dispute about the extent of the failure. The council accept that they failed to comply with (b) in that, although a notice was placed in the *Haringey Independent*, that newspaper did not circulate in the area of the site. They submit, however, that they complied with (a), in that they served notice, in accordance with the alternative in (a)(ii), on the adjoining owners and occupiers. Indeed, they went beyond that, by sending written notices to the occupiers of properties in the area even though, like the claimant, they were not strictly "adjoining" the application site. The claimant, on the other hand, contends that there was a failure to comply with (a) as well.

[26] The claimant's case on (a) is that the council had a discretion whether to proceed by way of displaying a site notice or by way of serving notices on adjoining owners or occupiers. Any statutory discretion has to be exercised in accordance with the spirit and purpose of the enabling statute and, in particular, for reasons relevant to the achievement of the statutory purpose: *R* v *Tower Hamlets London Borough Council, ex parte Chetnik Developments Ltd* [1988] AC 858, at pp872-873. The purpose of the publicity provisions is to secure notification of persons likely to be

[1] SI 1995 419

interested in the planning application. Consideration must, therefore, be given to whether that purpose is best achieved in an individual case by site notices, or by the service of notices on adjoining owners and occupiers. In an urban context, the former may be more appropriate, given that those likely to be interested will extend beyond those immediately adjoining the site. In the present case, however, the evidence is that the council simply followed their practice of serving notices on adjoining owners and occupiers, rather than using site notices. Accordingly, there was a failure to give proper consideration to the exercise of the statutory discretion.

[27] I have substantial doubts as to whether article 8(4)(a) imposes on a local planning authority an obligation to consider which of the two methods is best calculated to give notice of the application to those likely to be interested in the application. On the face of it, either of those methods is equally valid in every case: the relevant judgment has already been made by the Secretary of State, who, in making the 1995 Order, has formed the view that the purpose of ensuring that sufficient notice is given will be sufficiently achieved by a combination of (a) either of those methods plus (b) local advertisement.

[28] However, neither Mr Stephenson, for the council nor Mr Goatley, for Wolseley, has advanced that line of argument. Mr Stephenson's submission, adopted by Mr Goatley, is simply that the council's practice, which they followed in this case, was to serve notices on adjoining owners and occupiers, rather than to use site notices, and that the adoption of such a practice has not been shown to be unlawful.

[29] I am inclined to accept that, if and to the extent that it is incumbent on a local planning authority to make a discretionary judgment between the two alternatives in article 8(4)(a), it is permissible to decide to adopt a general practice rather than to assess each individual case separately, subject always to a willingness to consider any reasons advanced for departing from the practice in an individual case. It appears that the council had a practice of serving notices on adjoining owners and occupiers, and there is no suggestion that any reasons were advanced for departing from the practice in this case. So, I would reject the claimant's case based on a failure to consider which method to use in the individual case. There still remains a question whether the practice itself was lawful. The difficulty about that issue is the lack of evidence about the adoption of the practice or the reasons for it. The court has so little to go on that I would, if necessary, fall back on acceptance of Mr Stephenson's submission that the claimant has failed to establish that the practice followed in this case was unlawful.

[30] I do not consider, however, that a decision on the disputed issue under article 8(4)(a) is strictly necessary. In circumstances where the claimant did not otherwise receive notification of the planning application, the admitted failure to comply with article 8(4)(b) would have been a sufficient reason for quashing the grant of planning permission if there had been a timely challenge. If there was a failure to comply with article 8(4)(a) as well, it would be a point to put in the balance when considering the exercise of discretion to grant or withhold relief; but, in

my judgment, it would not carry much additional weight, and would not materially affect the outcome of the balancing exercise in the circumstances of the case as a whole. I, therefore, propose to say no more about it.

[31] I should make clear that, in concluding that the failure to comply with article 8(4)(b) would have been a sufficient reason for quashing the grant of planning permission, I am satisfied that the claimant was substantially prejudiced by the failure. He would have been in a position to make representations of substance in opposition to the proposed development, in respect of matters such as loss of amenity, visual intrusion, type of use (for example, the range of goods permitted to be sold by retail), hours of operation, traffic and car parking spaces. Such representations would have gone both to the principle of planning permission and to the conditions to be imposed if permission were granted. They might not have been successful, but they were of sufficient substance that he could legitimately complain of the denial of an opportunity to make them. The case for relief would have been reinforced by the fact that a substantial number of other residents would appear to have been unaware of the planning application, and would also have objected to it if they had been notified.

*Non-compliance with EIA requirements*

[32] The Town and Country Planning (Environmental Impact Assessment)(England and Wales) Regulations 1999, implementing Council Directive 85/337/EEC, make provision, in regulation 4, for a local planning authority to give a "screening opinion" as to whether a proposed development is an EIA development, that is, a development in respect of which an environmental statement is required because it is (a) a Schedule 1 development or (b) a Schedule 2 development likely to have significant effects on the environment by virtue of factors such as its nature, size or location. Requests for screening opinions are governed by regulation 5. Where a request is made under regulation 5(1), authorities are required, by regulation 5(4), to adopt a screening opinion within three weeks of receipt of the request, or such longer period as may be agreed with the person making the request.

[33] The regulation 5 procedures can also be engaged where an application for planning permission is made without an environmental statement. Regulation 7 provides:

> .–(1) Where it appears to the relevant planning authority that –
>
> (a) an application for planning permission which is before them for determination is a Schedule 1 application or Schedule 2 application; and
>
> (b) the development in question has not been the subject of a screening opinion or screening direction; and
>
> (c) the application is not accompanied by a statement referred to by the applicant as an environmental statement for the purposes of these Regulations,
>
> paragraphs (3) and (4) of regulation 5 shall apply as if the receipt or lodging of the application were a request made under regulation 5(1).

[34] It is common ground that the planning application in this case was a Schedule 2 application (as an urban development project exceeding

A 0.5ha, within para 10(b) of Schedule 2), that there had been no screening opinion or direction, and that the application was not accompanied by an environmental statement. The effect of regulation 7(1), read with regulation 5(4), was, therefore to require the council to adopt a screening opinion, which, in turn, required them to consider whether the development was likely to have significant effects on the environment by
B virtue of factors such as its nature, size or location. There was an admitted failure by the council to consider that question.

[35] Had that failure to comply with the EIA regulations been raised in a timely challenge, it would have led inevitably to the quashing of the planning permission in accordance with the principles laid down in *Berkeley* v *Secretary of State for the Environment, Transport and the Regions*
C *(No 1)* [2001] 2 AC 603[1]. The evidence of Mr Paul Tomkins, the council's development control manager (west), includes an expression of view that, had the matter been considered properly, an environmental statement would not have been required because there was no overall adverse environmental impact on this previously developed site. I am in no position to determine whether that is correct, but in any event it is irrelevant since
D *Berkeley* makes clear that individuals affected by the development have a directly enforceable right to have the need for an environmental statement considered before the grant of planning permission and that, where the question has not been considered, it is not normally permissible for the court to withhold relief in the exercise of its discretion.

[36] As explained below, however, the position with regard to the court's
E discretion is materially different where there has been undue delay.

*Delay: General principles*

[37] Section 31 of the Supreme Court Act 1981 provides:

> (6) Where the High Court considers that there has been undue delay in making an application for judicial review, the court may refuse to grant –
> (a) leave for the making of the application, or
> F (b) any relief sought on the application,
> if it considers that the granting of the relief sought would be likely to cause substantial hardship to, or substantially prejudice the rights of, any person or would be detrimental to good administration.
> (7) Subsection (6) is without prejudice to any enactment or rule of court which has the effect of limiting the time within which an application for
> G judicial review may be made.

[38] The effect of those provisions was explained in the speech of Lord Goff in *R* v *Dairy Produce Quota Tribunal for England and Wales, ex parte Caswell* [1990] 2 AC 738, at p747B-C:

> It follows that, when an application for leave to apply is not made
> H promptly and in any event within three months, the court may refuse leave on the ground of delay unless it considers that there is good reason for extending the period; but, even if it considers that there is such good reason, it may still refuse leave (or, where leave has been granted, substantive relief)

[1] [2000] 3 PLR 111

> if in its opinion the granting of the relief sought would be likely to cause hardship or prejudice (as specified in section 31(6)) or would be detrimental to good administration.

[39] In the present case, there was plainly undue delay. Elias J so found on the hearing of the permission application, I do not understand it to be in dispute before me, and, in any event, I am satisfied of the point. The central question for me is whether, although there were errors in the process leading to the grant of the planning permission, I should nonetheless refuse relief in the exercise of my discretion and, in particular, on the grounds of hardship, prejudice or detriment to good administration, as referred to in section 31(6). Although questions of hardship, prejudice and detriment to good administration were considered by Elias J at the permission stage, I need to form my own assessment on the basis of the evidence as a whole, and in the different context of the grant, or withholding of relief.

[40] I have mentioned that, on the basis of the decision in *Berkeley*, the court would normally have little, if any, discretion to withhold relief where planning permission has been granted in breach of the EIA requirements. Mr McCracken accepts that the full force of the strict approach laid down in *Berkeley* does not apply in a case of undue delay, though his precise position on this is not entirely clear to me. He does not contend that the court's discretion to withhold relief pursuant to section 31(6) of the 1981 Act is incompatible with general principles of EU law. At times, he appeared to accept that the court does have a real discretion in this case. But he also suggested that the refusal of relief would deny to the claimant an effective remedy for breach of his rights under the EIA directive, and would thereby conflict with the obligation of the court to ensure the effective protection of rights under EU law. The only authority he referred to on this subject was *Amministrazione delle Finanze dello Stato* v *San Giorgio SpA* C199/82 [1983] ECR 3595, and, even then, it was the briefest of references to paras 17-18 of the judgment, a passage which I find of no real assistance.

[41] For my part, I see no reason, in principle, why the exercise of discretion in accordance with section 31(6) of the 1981 Act should give rise to a problem under EU law. To entertain a claim outside the normal time limit but, at the same time, to take into account the consequences of that delay for third parties and good administration, does not render it impossible or excessively difficult to enforce EU rights, or involve any denial of effective protection, or otherwise offend general principles of EU law. For those reasons, and in the absence of any reasoned argument on this issue, I propose to approach the application of section 31(6) of the 1981 Act in a conventional way.

*Nature of the legal errors and the conduct of the council*

[42] The publicity requirements of the 1995 Order in relation to planning applications are of obvious importance, and the failure to comply with them caused substantial prejudice to the claimant. The requirements of the EIA Regulations are likewise of importance, as emphasised in *Berkeley*. Accordingly, if relief is to be refused on grounds of delay, factors of considerable weight have to be placed in the other side of the balance.

A [43] Apart from emphasising the importance of the errors they made, Mr McCracken makes various other criticisms of the council's conduct. In my judgment, none of them advances the claimant's case to any material extent. In particular, he seeks to make much of the fact that the planning officer who dealt with the planning application, and who prepared the report considered by the subcommittee in April 2000, was subsequently B dismissed for misconduct. The council states, however, that the misconduct related to a conflict of interest in an unrelated case, and did not touch on the planning application relevant to the present case. There is no basis for going behind that statement. For that reason, and because it was unnecessary for the fair determination of this case, I refused a very late application, made by Mr McCracken in the course of his substantive C submissions, for disclosure of the file relating to the officer's misconduct. In my view, such misconduct cannot provide a proper basis for adverse inferences in relation to the council's handling of the planning application at issue in this case. Further, although Mr McCracken has also drawn attention to the fact that the dismissal for misconduct was not disclosed to the court by the council in Mr Tomkins's first witness statement, in which D it was said that the individual concerned had "departed to travel round the world" and was not now contactable, I reject any suggestion that the council was deliberately misleading the court, or that this should count against the council when considering the question of relief.

[44] Nor do I think that anything turns on the council's shifting stance in response to the claimant's initial enquiries, and then his threat E to apply for judicial review and his commencement of proceedings: for example, the council's initial denial that a planning permission existed, followed by their discovery that there was a planning permission; or the gradual change of position regarding the adequacy of the newspaper advertisement or the receipt of a written notice by the claimant himself. Points can certainly be made about inefficiency and a reluctance to accept that errors had occurred, though the council's position was made more F difficult by the fact that the relevant planning officer had been dismissed, and most of the information had to be derived from perusal of the files. To my mind, however, what happened at this stage does not materially affect the exercise of discretion with regard to relief.

*Period of delay and the conduct of the claimant*

G [45] The delay in this case, a period of 32 months from the grant of planning permission to the commencement of proceedings, was, in any view, extremely long. On the other hand, it must be considered in conjunction with the explanation for the delay, namely the failure to comply with the publicity requirements, and the claimant's lack of knowledge of the planning application or of the grant of planning permission. The H claimant cannot fairly be criticised for failing to take action before he knew that there was anything to take action about.

[46] Mr Goatley has submitted, on behalf of Wolseley, that there were sufficient matters, individually or collectively, to put a reasonable person on enquiry long before mid-April 2003, when the claimant saw the activities on site that prompted him to act. There were the clearance

of trees and demolition of the coke office building in August 2002, and the erection of a hoarding in November 2002. In my recital of the factual background, I have referred to those matters and to the explanations given by the claimant (though it is right to note that he does not deal expressly with the demolition of the coke office building, which was noticed by Mrs Smith who lived opposite that building). In my judgment, the claimant's failure to appreciate a possible change in the planning status of the site, or to act earlier than he did, was understandable in the circumstances and should not count against him.

[47] Once the claimant did start to make enquiries, he acted with all reasonable speed. It was appropriate to make detailed investigations and to give careful consideration to the position before commencing legal proceedings in respect of a planning permission granted so long ago.

[48] Accordingly, leaving aside the significance of the delay itself, I see nothing in the claimant's conduct to put in the balance against the grant of relief.

*Hardship or prejudice to Wolseley*

[49] The central issue in the case is the hardship that would be suffered by Wolseley, or the prejudice that would be caused to Wolseley's interests (no distinction has been made in argument between hardship and prejudice), if the planning permission were quashed. Evidence on that is given in the witness statements of Mr Mason, Wolseley's development manager. In his first witness statement, made for the purposes of the permission hearing before Elias J, he refers to his own faxed letter, dated 1 April 2003, appointing Brennan as contractor for the works on site (in substitution for the contractor that had gone into liquidation in August 2002). The letter states:

> I am pleased to inform you that it is this company's intention to enter into a contract with Brennan Limited to complete the works for the above development. The Contract will be based on your Outline Proposal Presentation… The Contract Sum and Contractors Proposals will be as outlined in this document, the Fixed Price Contract being £1,850,000 (+VAT)…
>
> The Contract Programme will also be as detailed in the Outline Proposal Document, with works commencing 7th April 2003 and completing 37 calendar weeks thereafter. You should commence the detailed design works and the placing of subcontracts and ordering materials to facilitate the planned start on-site.
>
> Following successful negotiations we intend to send you a formal contract based on the terms and conditions of the JCT98 with Contractors Design…
>
> We reserve the right to withdraw the authority contained in this letter at any time and for any reason. If we do so, and the contract is not signed for any reason other than your failure to comply with the conditions set out in the tender documentation, then the company will reimburse you the reasonable and agreed costs wholly and necessarily incurred by you in properly proceeding with the works following your receipt of this letter.

[50] The programme of works was designed to be carried out in two phases, with phase 1 due for completion by the end of August 2003 and

A phase 2 by the end of December 2003. Brennan commenced work on site on 7 April 2003, and work has continued on site since then in line with the agreed programme, save that the works were approximately six weeks behind schedule as at the date of Mr Mason's third witness statement on 16 October.

[51] Although the letter of 1 April refers to the intention to conclude a
B formal contract, none has yet been signed. In his second witness statement, dated 25 July 2003, Mr Mason explained that the parties were, nevertheless, proceeding on the basis that the contract was governed by the terms and conditions referred to in the letter of 1 April.

[52] There was, in any event, a plain commitment on the part of Wolseley in respect of work done and materials ordered by Brennan
C in accordance with the terms of the letter of 1 April, and it is clear that termination of the works at any time would have led, or would now lead, to a financial claim by Brennan against Wolseley. It is neither necessary nor appropriate, in these proceedings, to determine the precise basis of such a claim – whether *quantum meruit*, as submitted by Mr McCracken, or a simple claim for breach of contract or even a tort-based claim, as
D Mr Goatley suggested by way of additional possibilities.

[53] Mr Mason says, in his first witness statement, that the position as at 11 June 2003 (this is the first date for which figures are given) was that the sum of £445,000 had been paid under stage payments, together with a further £8,000 in respect of professional fees. If Wolseley had terminated the "contract", an additional figure in the region of £650,000 would have
E been payable in respect of goods ordered. There might also have been a claim for loss of profits, but the terms of the letter of 1 April would have made such a claim very difficult to sustain, and I shall discount it for the purposes of my analysis.

[54] In his second witness statement, Mr Mason states that the position as at 23 July 2003 was that the sum of £584,250 plus VAT had been paid
F to the contractor, and a further sum of £213,973 had been valued by the contractor and agreed by the architect.

[55] In his third witness statement, Mr Mason refers to a letter dated 15 October 2003, in which the contractor indicated that an interim valuation would soon be issued in the sum of £1,426,116 plus VAT, and that it was currently committed, in terms of orders, to a figure of £1.8m overall. The
G sums will, no doubt, have become larger by the date when this judgment is handed down.

[56] Thus, the position is that, if the court quashes the planning permission, Wolseley will be at risk in respect of a total sum approaching £2m for work carried out and materials ordered in pursuance of an apparently valid planning permission. Wolseley's submission is that
H this amounts to hardship or prejudice amply sufficient to justify the withholding of relief.

[57] The claimant seeks to meet that case, or to reduce its force, in a number of ways.

[58] First, Mr McCracken submits that the court should confine its attention to loss that would have been incurred by Wolseley if it had

terminated the works on 15 April (the date of the claimant's first contact by telephone with Mr Mason), alternatively 30 April (the date of the claimant's first letter to Wolseley), alternatively 30 May (the date when the claim for judicial review was lodged). Wolseley ought to have stopped the works at one or other of those dates, in the light of the claimant's action in drawing attention to the invalidity of the planning permission, and threatening and then commencing the claim for judicial review. Mr McCracken complains about the absence of information about the costs that would have been incurred as at those various dates; but, since he articulated these specific alternative dates only in his oral submissions rather than in the claim form or skeleton argument, such a complaint cannot fairly be made. What can be said is that the costs would plainly have been lower than as at 11 June 2003, the first date for which detailed information is available, but would still have been large sums.

[59] Subject to later consideration of a separate point that the works were carried out in breach of condition and therefore unlawfully, I take the view that it was reasonable for Wolseley not to stop the works, even though it can be characterised as having taken a calculated commercial risk in proceeding as it did. It had an apparently valid planning permission. The time limit for a legal challenge to that permission had expired well over two years previously. It had had one false start in August 2002, with the contractor that went into liquidation. It had then entered into a legal commitment with Brennan on 1 April 2003, which, whatever its precise analysis, would expose it to a substantial claim if the works were stopped. In those circumstances, I do not think that a complaint about the validity of the permission or the threat of proceedings to challenge it were sufficient to make it unreasonable to continue with the works. No doubt, contractors are faced not infrequently with complaints about developments that do not mature into actual challenges. In the present case, moreover, the claimant's own explanation for the time spent in April and May before a claim for judicial review was lodged is that, especially in view of the lapse of time since the grant of planning permission, it was necessary to carry out detailed investigations and give careful consideration to whether a claim was justified. Wolseley cannot fairly be criticised for carrying on with the works while the claimant was considering his position.

[60] The lodging of the claim for judicial review made Wolseley's position more problematic. There was then a more clearly defined risk that the planning permission would be quashed, with the result that any work done under it would be unlawful. On the other hand, Wolseley had substantial grounds for resisting the grant of permission to apply for judicial review, and thereafter, when permission was granted by Elias J on 23 June, for resisting the grant of relief. At no time did the claimant seek an injunction to prevent the continuation of the works. This was entirely understandable, given that he would have been required to give a cross-undertaking in damages, and the sums involved were very large. But the absence of a cross-undertaking meant that any loss arising out of the cessation of the works would fall on Wolseley alone. Taking all those matters into account, it was, in my view, not unreasonable

of Wolseley to carry on with the works, even after the commencement of proceedings.

[61] It follows that I reject the claimant's submission that the costs incurred since, at the latest, the date when the claim for judicial review was lodged, should be left out of account when considering the hardship or prejudice that Wolseley would suffer if the planning permission were now quashed. But, even if I did confine my attention to the costs incurred as at the date when the claim for judicial review was commenced, the sums would still be large (precise figures are not given, but the costs as at 11 June are indicative), and the difference would not, in my view, affect the way in which the balance came down.

[62] The second main point that the claimant makes about Wolseley's case on hardship and prejudice is that Wolseley might, in practice, suffer no loss, or a loss far less than the cost of the works undertaken, even if the planning permission were quashed. That is because: (i) Wolseley might still be granted planning permission on its fresh application; and (ii) enforcement action might require steps other than complete demolition and removal of the works undertaken.

[63] In my view, the fresh application for planning permission cannot assist the claimant. The claimant is one of a substantial number of objectors who oppose the grant of planning permission. There is opposition to the development as a whole, not just to limited aspects of it. I cannot assess whether permission is likely to be granted, whether by the council or by the Secretary of State on appeal. In any event, if permission were likely to be granted, that might be thought to weaken, rather than strengthen, the case for quashing the existing permission. I must also take account of the evidence that Wolseley would have to incur substantial (though unquantified) retendering costs if it were required to stop the works now, on the quashing of the existing permission, but were then able to proceed with the balance of the works, on the grant of permission on the fresh application.

[64] Nor does the point on enforcement action avail the claimant. It would be a matter of judgment for the council how to proceed with regard to enforcement. The powers conferred by section 173 of the Town and Country Planning Act 1990 would give it considerable flexibility. It is possible that something less, even far less, than demolition and removal of the entire development would be required. But all this is speculative, and it is not possible to form any measured assessment of the actual outcome. The fact is that, if the planning permission were quashed and no new permission were forthcoming, the entire development would be unlawful and Wolseley would be at risk of being required to demolish and remove it all. It is a risk that cannot be dismissed as insignificant.

[65] A third point advanced by the claimant in relation to the financial consequences for Wolseley is that it would have a claim in negligence against the council in respect of any financial loss that it suffered as a result of the council's non-compliance with the publicity requirements and EIA regulations. I accept that Wolseley might have such a claim, but its strength is not something that has been investigated in any detail before

me, and I do not think that I should attach much weight to the speculative possibility of Wolseley recouping its losses from the council. A

[66] In addition to the above matters, the claimant seeks to counter the force of Wolseley's case on hardship or prejudice by further submissions concerning Wolseley's conduct. I think it more convenient to consider those submissions under a separate heading.

B

*Wolseley's conduct*

[67] The claimant contends that: (i) Wolseley was the author of its own misfortune by failing to check on whether the council complied with the publicity requirements and the EIA requirements; and (ii) Wolseley has acted unlawfully in carrying out the works in breach of certain of the conditions of the planning permission. Both matters, it is submitted, should weigh heavily in the balance against the withholding of relief on grounds of hardship and prejudice to Wolseley. C

[68] As to the first matter, Mr McCracken submits that Wolseley, a substantial business with professional advisers, should have taken steps to ensure that the council had complied with the relevant requirements. This would have been an elementary precaution, given the risks to Wolseley if the council fell into error. It would have been simple to check whether a site notice had been displayed (ie the form of publicity that the claimant says ought to have been used), and whether a screening opinion had been placed on the register (as should have been done if one had been adopted). Having failed to take such steps, Wolseley should not be allowed to rely on hardship or prejudice suffered by its reliance on a planning permission that was granted in non-compliance with the relevant requirements and was therefore unlawful. D E

[69] If that argument were accepted, it seems to me that it would be tantamount to saying that a developer is under an obligation to monitor the lawfulness of the steps taken by a local planning authority at each stage of its consideration of a planning application. In my judgment, it would be wrong to go down that line. It is not warranted by the legislative scheme, which places the relevant responsibilities on the local planning authority; and it would give rise to practical difficulties if applicants were required, at each stage, to check on the authority's discharge of their responsibilities. Applicants for planning permission are entitled to rely on the local planning authority to discharge the responsibilities placed upon them. They should not be held accountable for the authority's failure to comply with relevant requirements, at least where, as here, they cannot be said to have caused or contributed to that failure by their own conduct. In that respect, I see no distinction of principle between a private individual acting for himself and a substantial developer with professional advisers. F G

[70] The second point to be considered under this heading concerns Wolseley's failure to comply with the conditions of the existing planning permission. It is common ground that some of those conditions had not been complied with, as they should have been, before commencement of the works, and that some have not been complied with even now. It follows, submits Mr McCracken, that Wolseley has acted unlawfully in carrying out the works and does not come to the court with H

"clean hands"; and although this is not relied upon as an automatic bar to Wolseley's reliance on hardship and prejudice, it is said that it should weigh heavily against Wolseley in a case where Wolseley is seeking an exercise of discretion in its favour.

[71] The principal conditions relied on are these:

> 3. Samples of all materials to be used for the external surfaces of the development shall be submitted to, and approved in writing by, the Local Planning Authority before any development is commenced...
>
> ...
>
> 5. Notwithstanding the details of landscaping referred to in the application, a scheme for the landscaping and treatment of the surroundings of the proposed development to include detailed drawings of:
>
> a. those existing trees to be retained
>
> b. those existing trees to be removed
>
> c. those existing trees which require thinning, pruning, pollarding or lopping as a result of this consent. All such work to be agreed with the Council's Arboriculturalist
>
> d. those new trees and shrubs to be planted together with a schedule of species
>
> shall be submitted to, and approved in writing by, the Local Planning Authority prior to the commencement of the development...
>
> 6. Details of a scheme depicting those areas to be treated by means of hard landscaping shall be submitted to, approved in writing by, and implemented in accordance with the approved details...
>
> 7. The existing trees on the site shall not be lopped, felled or otherwise affected in any way... without the prior written permission of the Local Planning Authority...
>
> ...
>
> 10 Details of the proposed foundations in connection with the development hereby approved and any excavation for services shall be agreed with the Local Planning Authority prior to the commencement of the building works.

[72] It appears that none of those conditions had been fulfilled at the date of commencement of the works or, indeed, when the works were continued following the lodging of the claim for judicial review. The matters were the subject of discussion, but nothing had been finalised. Written confirmation of the acceptability of materials under condition 3 was given only on 12 August 2003. Drawings relating to condition 5 were still being considered on 16 October, the date of Mr Mason's most recent witness statement. A supplementary letter in respect of condition 6 was still outstanding as at that date. There had been an admitted failure to comply with condition 7, in that an area of silver birches on the southern part of the site had been felled before the commencement of the main works, and without prior approval (though it would seem that their removal had been shown on the drawings); and there had been no formal compliance with the condition as at 16 October, though the evidence is that the position with regard to various trees that remained on site had been discussed and agreed with the council's arboriculturalist on site in May. Certain details under condition 10 were also still outstanding as at 16 October.

[73] It is right to note that the felling of trees without prior approval was one of the matters of particular concern to the claimant, and that the need for conditions requiring the implementation of strict landscaping and tree planting schemes had been highlighted in the officer's report to the subcommittee that had granted the planning permission in April 2000. A

[74] What is the consequence of those failures to comply with conditions prior to the commencement of the works? Technically, it rendered the works unlawful and open to enforcement action. For the council and Wolseley, however, Mr Stephenson and Mr Goatley submit that the court should have regard to the realities of the matter. Developers not infrequently commence works without full compliance with preconditions, whether through mistake, misunderstanding or calculated commercial decision. The local planning authority are responsible for monitoring the situation and deciding whether to use the enforcement powers available to them. In this case, the council, although annoyed that prior approval had not been sought for the felling of trees on the southern part of the site, have not been unduly concerned with the situation overall and have not seen fit to intervene to stop the works. Discussions have continued over compliance with the substance of the conditions, and there is no reason to doubt that there will be a satisfactory resolution of the outstanding issues. B C D

[75] It is further submitted, on behalf of the council and Wolseley that, although the authorities establish the principle that operations carried out in breach of condition cannot be relied upon as a lawful commencement of the development and can be enforced against, they also indicate the need for a degree of realism. Thus, the principle must be applied with common sense and with due regard to the facts of the particular, so that there may be a finding of lawful commencement where the substance of a condition has been complied with, even if there has been a technical failing: *R* v *Flintshire County Council, ex parte Somerfield Stores Ltd* [1998] PLCR 336, at pp351F-352. And if enforcement action is taken, when the merits of the enforcement proceedings come to be considered, it is necessary to take into account the situation as it exists at that time and, in particular, whether or not, at that time, any approval required by condition has been obtained: *FG Whitley & Sons Co Ltd* v *Secretary of State for Wales* (1992) 64 P&CR 296[1]. E F

[76] I have not found it easy to decide, in the light of those rival submissions, what importance to attach to the fact that the works were undertaken here in breach of the conditions of the planning permission and, therefore, unlawfully. It undermines Wolseley's argument that it acted lawfully throughout, in implementation of an apparently lawful planning permission, and it puts Wolseley in a generally less attractive position when seeking an exercise of discretion in its favour. On the other hand, I think it right to take account of the fact that what has happened is not exceptional in practice and that the enforcement authority, the council, have not seen fit to intervene. Moreover, the indications are that G H

[1] [1992] 3 PLR 72

the required approvals, if not already now given, will be forthcoming before long end, as a result of continuing discussion over the details. So, the breaches of condition relate, essentially, to timing rather than to substance.

[77] In all the circumstances, I have concluded that the point should count against Wolseley when deciding whether to withhold relief on the ground of detriment or prejudice to Wolseley, but that it should not be given the degree of weight that Mr McCracken seeks to attribute to it. I do not regard it as negating my finding that it was reasonable for Wolseley to continue with the works after the claimant had questioned the validity of the planning permission, or after he had threatened and then commenced proceedings for judicial review.

*Detriment to good administration*

[78] Both Mr Stephenson for the council and Mr Goatley for Wolseley place considerable reliance on detriment to good administration as a further ground for withholding relief. They point out that the refusal of relief on this ground was upheld by the House of Lords in *Caswell*. In that case, Lord Goff, having referred to observations of Lord Diplock in *O'Reilly* v *Mackman* [1983] 2 AC 237 on the public interest in good administration, went on at pp749F-750B:

> I do not consider that it would be wise to attempt to formulate any precise definition or description of what constitutes detriment to good administration. This is because applications for judicial review may occur in many different situations, and the need for finality may be greater in one context than in another. But it is of importance to observe that section 31(6) recognises that there is an interest in good administration independently of hardship, or prejudice to the rights of third parties, and that the harm suffered by the applicant by reason of the decision which has been impugned is a matter which can be taken into account by the court when deciding whether or not to exercise its discretion under section 31(6) to refuse the relief sought by the applicant. In asking the question whether the grant of such relief would be detrimental to good administration, the court is at that stage looking at the interest in good administration independently of matters such as these. In the present context that interest lies essentially in a regular flow of consistent decisions, made and published with reasonable dispatch; in citizens knowing where they stand, and how they can order their affairs in the light of the relevant decision. Matters of particular importance, apart from the length of time itself, will be the extent of the effect of the relevant decision, and the impact which would be felt if it were to be re-opened. In the present case, the court was concerned with a decision to allocate part of a finite amount of quota, and with circumstances in which a re-opening of the decision would lead to other applications to re-open similar decisions which, if successful, would lead to re-opening the allocation of quota over a number of years. To me it is plain, as it was to the judge and to the Court of Appeal, that to grant the appellants the relief they sought in the present case, after such a lapse of time had occurred, would be detrimental to good administration...

[79] It is submitted that, in the planning context, there is a particular need for prompt challenges and certainty, in the interests of good

administration. The need to bring any challenge to a planning permission speedily has been emphasised repeatedly in the cases (although they must now be read subject to the qualifications in *R (on the application of Burkett)* v *Hammersmith and Fulham London Borough Council* [2002] UKHL 23; [2002] 1 WLR 1593[1]). Mr Goatley cites, by way of example, the reasons given by Pill LJ in *R* v *Newbury District Council, ex parte Chieveley Parish Council* (Court of Appeal judgment dated 23 July 1992[2]): A B

> A reason for that approach is that a planning permission is contained in a public document which potentially confers benefit on the land to which it relates. Important decisions may be taken by public bodies and private bodies and individuals upon the strength of it, both in relation to the land itself and in the neighbourhood. A chain of events may be set in motion. It is important to good administration that, once granted, a permission should not readily be invalidated. As confirmed in the House of Lords, section 31(6) recognises that there is an interest in good administration independent of hardship, or prejudice to the rights of third parties. The court is entitled to look at the interest in good administration independently of those other matters. It is important that citizens know where they stand and how they can order their affairs in the light of the relevant decision (*Caswell*…). In my judgment, weight should be given to this aspect of the case notwithstanding the absence of convincing evidence that the applicants for planning permission have been prejudiced by the delay… I have no doubt that interests of good administration, which, as contemplated by *Caswell*, extend beyond the interests of the parties to the litigation, should constitute an important factor in the decision. C D

[80] Although Mr Goatley also relies on *R* v *North West Leicestershire District Council, ex parte Moses (Costs)* unreported 12 April 2000, a decision of the Court of Appeal, which cites *Chieveley*, it was based on very different facts, and I do not think that it adds to the relevant principles. E

[81] Mr McCracken, on the other hand, relies upon the following passage in the judgment of the Court of Appeal, *per* Sedley LJ, in *R (on the application of Lichfield Securities Ltd)* v *Lichfield District Council* [2001] EWCA Civ 304; [2001] PLCR 32[3], at pp539-540, submitting that, although not an essential part of the court's reasoning in the case, it is of strong persuasive authority: F

> 39. The question of possible detriment to good administration arises under section 31(6) only if there has been undue delay. Mr Mole, for LDC, has laid understandable stress on this ground for denying relief which is otherwise called for. It is a relatively unexplored ground, if one may judge by its brief appearance in Fordham's encyclopaedic *Judicial Review Handbook* (2nd ed), para 26.9.3, no doubt partly for the reasons indicated in Lord Goff's speech in *R* v *Dairy Produce Quota Tribunal for England and Wales, ex parte Caswell* [1990] 2 AC 738 at pp749-750. Lord Goff was careful to avoid a formulaic approach, limiting himself to the specific effect in that case of a very long delay on the desirability of a regular flow of consistent G H

[1] [2002] 2 PLR 90
[2] (1998) 10 Admin LR 676
[3] [2001] 3 PLR 33

> decisions by the tribunal in question. But a further reason for the relative infrequency of decisions based upon good administration is, in our view, that it can come into play only: (a) where undue delay has occurred; and (b) – in practice – where the consequent hardship or prejudice to others is insufficient by itself to cause relief to be refused. In such a situation it can rarely, if ever, be in the interests of good administration to leave an abuse of public power uncorrected. Indeed Fordham records the decision of May J in *R* v *Mid-Warwickshire Licensing Justices, ex parte Patel* [1994] COD 251 that, despite undue delay, the interests of good administration were served not by withholding but by granting relief.

[82] In my judgment, it is plain that detriment to good administration is capable, in principle, of amounting to a sufficient reason for withholding relief. Section 31(6) of the 1981 Act so contemplates, and *Caswell* shows that circumstances can arise where relief may properly be refused on the ground of detriment to good administration, even in the absence of proven hardship or prejudice to third parties. I do not think that the Court of Appeal in *Lichfield* can have had in mind the kind of situation that arose in *Caswell* when making the observations that it did about the limited room for reliance on detriment to good administration. The context in *Caswell* was, however, very different from that of the present case, and the observations of the Court of Appeal have much greater relevance to the present context. But even then they should, in my view, be read not as precluding the refusal of relief on the ground of detriment to good administration, but as serving to emphasise the need for caution in deciding whether the grant of relief really would be detrimental to good administration and, if so, how much weight to attach to that detriment.

[83] I do not doubt the importance of certainty in the context of planning decisions, for reasons of the kind mentioned in *Chieveley*. Third parties are entitled to rely, and do in practice rely, on the information contained in the planning register, and to quash a planning decision long after it was made will undermine the basis upon which people have acted in the meantime. The developer that undertakes work in reliance on the permission is likely to be the person principally affected, though is also likely to be the person best placed to establish substantial hardship or prejudice. But it would be wrong to focus on the developer alone. Others may also have relied on the planning permission and have ordered their affairs accordingly, eg in negotiating the price of property near the development. It is very unlikely that all those affected could be identified or that specific hardship or prejudice could be proved in relation to each. Nevertheless, it is contrary to the interests of good administration to undermine the basis upon which they have acted (and at the same time to create uncertainty as to the reliance that can safely be placed on apparently valid planning permissions in the future). I therefore consider that detriment to good administration ought to be taken into account as a separate and additional factor relevant to the exercise of discretion to quash. But it is of only secondary significance, compared with the hardship or prejudice to the developer.

[84] In reaching that conclusion, I have borne in mind that the interests of good administration cut both ways, in that they are also served by correcting legal errors where they have occurred. But, in my view, there

would still be a net detriment to good administration if the planning A
permission were quashed so long after it was granted.

*Convention issues*

[85] The points raised in relation to the Convention have been touched
on only very briefly in argument. Mr McCracken submits that Article 8
is engaged, given the impact of the development on the claimant's home, B
but does not contend that refusal of the relief sought would amount to an
infringement of Article 8. He also submits that Article 6 is engaged and that
refusal of relief in the circumstances of the case would amount to a denial
of a practical and effective opportunity to challenge the grant of planning
permission; and he cites, in that context, a decision of the European Court
of Human Rights in *De Geouffre de la Pradelle* v *France* [1992] A253-B. But C
I confess that I have found the argument under Article 6 difficult to follow,
and am wholly unpersuaded that there is anything in it.

[86] Wolseley also relies upon the Convention. It is contended in
Mr Goatley's skeleton argument that to grant relief after such a lapse of
time would amount to a breach of Wolesely's right under Article 6 to a
hearing within a reasonable time and would give rise to the removal of, or D
a disproportionate interference with, the claimant's property rights under
Article 1 of the First Protocol. The contentions were not elaborated orally
and, again, I am not persuaded that there is any force in them.

[87] Thus, in my judgment, the resolution of this case is not materially
affected by the Convention.

*Discussion and conclusion* E

[88] I have set out the various factors referred to in argument as being
relevant to the exercise of the court's discretion and I have indicated my
views on each of them. It remains for me to determine whether the overall
balance favours the grant or withholding of relief.

[89] In short, Mr McCracken submits that the court should act to
protect the claimant's position, by quashing the planning permission on F
the grounds of the council's serious errors and enabling the claimant to
advance the objections to the development that he was previously denied
the opportunity to make. Once the claimant took issue with the validity
of the planning permission, Wolseley took a calculated risk to proceed
with the work, which should not deprive the claimant of a remedy. The
claimant's interest in his home should prevail over money. For the council G
and Wolseley, by contrast, it is submitted that relief should be withheld
because of the financial consequences for Wolseley and the detriment to
good administration if the planning permission were quashed.

[90] The conclusion I have reached is that I should refuse an order
quashing the planning permission. It is unnecessary to repeat what
I have said about each of the relevant factors. I stress that, in the claimant's H
favour, I attach substantial weight to the fact that the failure to comply
with the publicity requirements and the EIA requirements were serious
procedural errors, and that the claimant has been denied an opportunity
to make representations in opposition to a development that affects his
home. I also bear in mind that the claimant is not to blame for the delay.

But there has been undue delay within section 31(6) and it has been a very long delay; and to quash the planning permission after that lapse of time, and in the circumstances now existing would, in my judgment, cause very substantial hardship or prejudice to Wolseley. The adverse financial consequences for Wolseley, even taken at their lowest, are very large and are not to be discounted by reference to the speculative possibility that the fresh application for planning permission might succeed, or that full enforcement action might not be taken, or that losses might be recouped by a claim in damages against the council. Even allowing for the criticisms of Wolseley's conduct, to the extent that I have accepted them, I take the view that the hardship or prejudice to Wolseley is a sufficient reason for the refusal of a quashing order. To grant such an order would also be detrimental to good administration, but, in the event, I do not need to rely on this as a factor tipping the balance in favour of refusal.

[91] The same considerations against the grant of relief do not apply to the declaration sought by the claimant as an alternative to a quashing order. To declare that the council failed to comply with the relevant publicity requirements and EIA requirements would serve to underline the council's failings and would provide some satisfaction to the claimant, but without affecting the validity of the planning permission itself or, therefore, of works carried out pursuant to it. It may not be strictly necessary, since this judgment can speak for itself, but I think it appropriate in all the circumstances to grant such a declaration.

[92] In his judgment on the permission application, Elias J said that "[i]n substance it seems to me I have to try and determine where the lesser injustice is caused": see para 20. The court is, in my view, engaged in a similar exercise at the substantive stage, within the framework of section 31(6). In my judgment, the outcome in this case that produces the lesser injustice is that the claimant should succeed to the extent of obtaining declaratory relief but that a quashing order should be refused.

[93] I will hear counsel as to the precise form of the declaration and as to any consequential matters if the parties are unable to agree on a form of order in the light of this judgment.

*Declaratory relief awarded.*

# R (on the application of Beresford) *v* Sunderland City Council

[2003] UKHL 60

HOUSE OF LORDS

*Lord Bingham of Cornhill, Lord Hutton, Lord Scott of Foscote, Lord Rodger of Earlsferry and Lord Walker of Gestingthorpe*

13 November 2003

**Town or village green – Section 22 of Commons Registration Act 1965 – Registration – Sports and pastimes indulged in for 20 years – User tolerated and encouraged by owner – Whether user as of right – Whether user with implied permission – Whether implied permission will always negate user as of right**

The respondent council were the owners of a 10 acre area of grassed arena (the sports arena), which had been created in connection with the building of Washington new town. Rows of wooden benches and a single hard cricket pitch were installed and maintained and the council, the statutory successors in title to several public authorities, regularly cut the grass. In 1998, the council obtained planning permission to develop the sports arena as part of a college of further education. The appellant, on behalf of a number of local inhabitants, claimed that, from 1977, the land had been used for various recreational activities – from team games to the walking of dogs. He therefore applied to have it registered as a town or village green under section 13 of the Commons Registration Act 1965. The application was refused on the ground that the use had not been "as of right" within the meaning of section 22 of the 1965 Act. The appellant's application for judicial review of that decision was also refused; Smith J concluded that the user was with implied permission, and therefore not as of right: see [2001] EWCA Civ 1218; [2001] 1 WLR 1327. On appeal from a decision of the Court of Appeal ([2001] 3 PLR 120) dismissing the appellant's appeal, it was accepted that the sports arena had been used for "lawful sports and pastimes" by inhabitants of a locality for 20 years, the council contending that this was with implied permission; the sole issue was whether the use had been "as of right".

***Held*** The appeal was allowed; the sports arena should be registered as a town or village green. The user was as of right within the meaning of section 22(1) of the Commons Registration Act 1965.

***Per Lord Scott of Foscote***: The various statutory provisions relating to the acquisition and ownership of open space or recreational land were relevant as background against which the implications of the recreational use of the sports arena for the period 1977 to 1999 (when the decision had been made to use the land for educational purposes) ought to be assessed. The fact that the land was publicly owned was highly material. The public bodies that owned the sports arena were discharging functions for the

A public, and the provision of the benches and the mowing of the grass were not indicative of a precatory permission but of a public authority, mindful of its public responsibilities and function, desirous of providing recreational facilities to the inhabitants of the locality; the inhabitants were entitled to believe that they had the right to use the sports arena on a permanent basis. The nature of the implied permission from the public
B authority owners was not such as to prevent the use being "as of right". Although an implied permission will, in most cases, be inconsistent with use as of right, the conclusion is an evidentiary one and not a rule of law. In the present case, that was not a correct evidentiary conclusion.

***Per Lord Walker of Gestingthorpe***: In relation to town or village greens, it would be wrong to treat a landowner's silent and passive acquiescence
C in persons using his land as having the same effect as permission communicated to those persons. There was no evidence of overt acts on the part of the council, or its predecessors, to justify the conclusion of an implied licence; the provision of benches, a single hard cricket pitch and grass cutting could not be regarded as overt acts communicating permission.

D

**Cases referred to in the opinions**


*Attorney-General* v *Poole Corporation* [1938] Ch 23; [1937] 3 All ER 608

*Bridges* v *Mees* [1957] Ch 475; [1957] 3 WLR 215; [1957] 2 All ER 577

*Burrows* v *Lang* [1901] 2 Ch 502

*Cumbernauld and Kilsyth District Council* v *Dollar Land (Cumbernauld) Ltd* 1993 SC (HL) 44, HL; 1992 SLT 1035, Ct Sess

E *Dalton* v *Henry Angus & Co; Commissioners of HM Works and Public Buildings* v *Henry Angus & Co; sub nom Angus & Co* v *Dalton* [1881] All ER 1; (1881) LR 6 App Cas 740

*Davies* v *Du Paver* [1953] 1 QB 184; [1952] 2 All ER 991; (1952) 2 TLR 890, CA

*ER Ives Investment Ltd* v *High* [1967] 2 QB 379; [1967] 2 WLR 789; [1967] 1 All ER 504, CA

*Folkestone Corporation* v *Brockman* [1914] AC 338

F *Gardner* v *Hodgson's Kingston Brewery Co Ltd* [1903] AC 229

*Hall* v *Beckenham Corporation* [1949] 1 KB 716; [1949] 1 All ER 423; 65 TLR 146; 47 LGR 291

*Herrington* v *British Railways Board* [1972] AC 877; [1972] 2 WLR 537; [1972] 1 All ER 749; [1972] EGD 619; (1972) 223 EG 939, HL

*Jones* v *Bates* [1938] 2 All ER 237

*Mann* v *Brodie* (1885) 10 App Cas 378

G *Marquis of Bute* v *McKirdy & McMillan Ltd* 1937 SC 93

*Mills* v *Silver* [1991] Ch 271; [1991] 2 WLR 324; [1991] 1 All ER 449; (1990) 61 P&CR 366, CA

*Napier's Trustees* v *Morrison* (1851) 13 D 1404

*R (on the application of Beresford)* v *Sunderland City Council; sub nom R* v *Sunderland City Council, ex parte Beresford* [2001] EWCA Civ 1218; [2002] QB 874; [2002] 2 WLR 693; [2001] 4 All ER 565; [2002] 1 P&CR 32; [2001] 3 PLR 120; [2001] JPL
H 1307, CA; [2001] 1 WLR 1327; (2000) 82 P&CR 19, AC

*R* v *Oxfordshire County Council, ex parte Sunningwell Parish Council* [2000] 1 AC 335; [1999] 3 WLR 160; [1999] 3 All ER 385; (2000) 79 P&CR 199; [1999] 2 EGLR 94; [1999] 31 EG 85; [2000] JPL 384, HL

*R* v *Suffolk County Council, ex parte Steed* (1996) 75 P&CR 102; [1997] 1 EGLR 131; [1997] 10 EG 146, CA

*Scottish Property Investment Co Building Society* v *Horne* (1881) 8 R 737 A
*Sturges* v *Bridgman* [1879] 11 ChD 852

**Appeal against the decision of the Court of Appeal**

This was an appeal by the appellant, Mrs Pamela Beresford, from the Court of Appeal's dismissal of her appeal from a decision of Smith J refusing her application for judicial review of a decision of the respondents, Sunderland B
City Council, not to register an area of land as a town or village green under section 13 of the Commons Registration Act 1965.

*George Laurence QC and Douglas Edwards (instructed by Southern Stewart & Walker, of Sunderland) appeared for the appellant, Pamela Beresford.*

*Philip Petchey (instructed by the solicitor to the Sunderland City Council)* C
*represented the respondent council.*

*The following opinions were delivered.*

**LORD BINGHAM**:

My lords,

[1] The issue in this appeal is whether Sunderland City Council erred D
in law in refusing to register as a "town or village green", under the Commons Registration Act 1965, an area of land known as the sports arena (the land) situated close to the town centre of Washington, Tyne and Wear. I am indebted to my noble and learned friends, Lord Scott of Foscote, Lord Rodger of Earlsferry and Lord Walker of Gestingthorpe, for their summaries of the relevant facts and the history of these proceedings, E
which I gratefully adopt and need not repeat.

[2] As defined in section 22 of the 1965 Act, before its amendment by section 98 of the Countryside and Rights of Way Act 2000, the expression "town or village green" means (for present purposes):

> land... on which the inhabitants of any locality have... indulged in [lawful] F
> sports and pastimes... as of right for not less than twenty years.

As Pill LJ rightly pointed out in *R* v *Suffolk County Council, ex parte Steed* (1996) 75 P&CR 102, at p111:

> it is no trivial matter for a landowner to have land, whether in public or private ownership, registered as a town green... G

It is accordingly necessary that all the ingredients of this definition should be met before land is registered, and decision makers must consider carefully whether the land in question has been used by the inhabitants of a locality for indulgence in what are properly to be regarded as lawful sports and pastimes and whether the temporal limit of 20 years' or more indulgence has been met. These ingredients of the definition can give rise H
to contentious and difficult questions. But they do not do so in this case. The only difference between the parties, upon which the appeal turns, is whether the admitted use of the land by the inhabitants of the locality for indulgence in lawful sports and pastimes for not less than 20 years was "as of right".

A [3] In this context, it is plain that "as of right" does not require that
the inhabitants should have a legal right, since in this, as in other cases
of prescription, the question is whether a party who lacks a legal right
has acquired one by user for a stipulated period. It is also plain that "as
of right" does not require that the inhabitants should believe themselves
to have a legal right: the House so held in *R* v *Oxfordshire County Council,*
B *ex parte Sunningwell Parish Council* [2000] 1 AC 335, at pp354 and 356.
It is clear law, as summarised in the last-mentioned decision, that, for
prescription purposes under the Prescription Act 1832, the Rights of Way
Act 1932 and the 1965 Act, "as of right" means *nec vi, nec clam, nec precario,*
that is, "not by force, nor stealth, nor the licence of the owner": see pp350,
351 and 353-354. In this case, there was no question of force or stealth.
C So the only question is whether the inhabitants' user was by the licence of
the owner.

[4] It was not suggested that the council had expressly licensed the
inhabitants' use of the land, either in writing or orally. The argument was
accordingly directed to whether it was ever possible to imply a licence by
a landowner to use land in the manner prescribed by the statute, and, if
D so, whether the facts here could properly be held to give rise to such an
implication.

[5] I can see no objection in principle to the implication of a licence
where the facts warrant such an implication. To deny this possibility
would, I think, be unduly old-fashioned, formalistic and restrictive.
A landowner may so conduct himself as to make it clear, even in the
E absence of any express statement, notice or record, that the inhabitants' use
of the land is pursuant to his permission. This may be done, for example,
by excluding the inhabitants when the landowner wishes to use the land
for his own purposes, or by excluding the inhabitants on occasional days:
the landowner, in this way, asserts his right to exclude, and so makes it
plain that the inhabitants' use on other occasions occurs because he does
F not choose, on those occasions, to exercise his right to exclude and so
permits such use.

[6] Authority, however, establishes that a licence to use land cannot be
implied from the mere inaction of a landowner with knowledge of the use
to which his land is being put. In *Davies* v *Du Paver* [1953] 1 QB 184, which
concerned a private right, Morris LJ said, at p210:

G

> Before Mr Davies could establish a claim based on prescription the evidence would have to show that the owner of the servient tenement had knowledge of what was happening, or as an ordinary owner must be taken to have had reasonable opportunity of knowledge, and that, having power to prevent it, he did not intervene.

H In *Mills* v *Silver* [1991] Ch 271, which also concerned a private right,
Dillon LJ acknowledged, at pp279G-280A, that:

> it would be easy to say... that there is an established principle of law that no prescriptive right can be acquired if the user by the dominant owner of the servient tenement in the particular manner for the appropriate number of years has been tolerated without objection by the servient owner. But

> there cannot be any such principle of law because it is, with rights of way, fundamentally inconsistent with the whole notion of acquisition of rights by prescription. It is difficult to see how, if there is such a principle, there could ever be a prescriptive right of way. A

At p281G, Dillon LJ added:

> It is to be noted that a prescriptive right arises where there has been user as of right in which the servient owner has, with the requisite degree of knowledge... acquiesced. Therefore mere acquiescence in or tolerance of the user by the servient owner cannot prevent the user being user as of right for purposes of prescription. B

Parker LJ, at p290C, was of the same opinion: C

> The true approach is to determine the character of the acts of user or enjoyment relied on. If they are sufficient to amount to an assertion of a continuous right, continue for the requisite period, are actually or presumptively known to the owner of the servient tenement and such owner does nothing that is sufficient...
>
> ...
>
> I add only this, that any statement that the enjoyment must be against the will of the servient owner cannot mean more than "without objection by the servient owner." If it did, a claimant would have to prove that the right was contested and thereby defeat his own claim. D

In *Sunningwell* [2001] 1 AC 335, it was held by the House that the landowner's toleration of the local inhabitants' user of the land in question was not inconsistent with such user having been as of right, E and so did not prevent registration of the land in question as a town or village green. As my noble and learned friends Lord Rodger and Lord Walker point out, some caution is required of English lawyers reading the Scottish authorities, since the applicable legislation is not the same, and "tolerance" is used to mean not "acquiescence" but "permission". F It does, however, appear that the Scottish approach to prescription, as applied to public rights of way, is close to that of the English. As the Lord President (Hope) put it in *Cumbernauld and Kilsyth District Council* v *Dollar Land (Cumbernauld) Ltd* 1992 SLT 1035, at p1041G, in a passage expressly approved by the House of Lords (1993 SC (HL) 44, at p47):

> where the user is of such amount and in such manner as would reasonably be regarded as being the assertion of a public right, the owner cannot stand by and ask that his inaction be ascribed to his good nature or to tolerance. If his position is to be that the user is by his leave and licence, he must do something to make the public aware of that fact so that they know that the route is being used by them only with his permission and not as of right. G

[7] Recognising that the authorities preclude reliance upon mere inaction as giving rise to an implied licence to use the land, the council have placed reliance upon their conduct in mowing the grass on the land and providing benches for the accommodation of spectators and other users of the land. This, it was said, showed that the council were encouraging the public to use the land, from which their licence to do H

so could be implied. Both the mowing of the grass and the provision of benches are open to more than one explanation. But the argument is, in my opinion, open to a more fundamental objection. As already pointed out, the 1965 Act drew heavily upon principles established under the Acts of 1832 and 1932, relating to private and public rights of way respectively, and in neither of these instances could acts of encouragement by the servient owner be relied upon to contend that the user by the dominant owner had not been as of right. Such conduct would indeed strengthen the hand of the dominant owner. Here, the conduct is, in any event, equivocal: if the land were registered as a town or village green, so enabling the public to resort to it in exercise of a legal right and without the need for any licence, one would expect the council to mow the grass and provide some facilities for those so resorting, thus encouraging public use of this valuable local amenity. It is hard to see how the self-same conduct can be treated as indicating that the public had no legal right to use the land and did so only by virtue of the council's licence.

[8] In the decision under challenge, the council considered that there was evidence, which they accepted, of an implied licence, thus enabling the inference to be drawn that the use by local inhabitants for statutory purposes had not been as of right. In her clear and helpful judgment ([2001] 1 WLR 1327) Janet Smith J accepted that conclusion: see pp1340-1341. For the reasons given by Dyson LJ, with which Latham LJ and Wilson J agreed ([2001] EWCA Civ 1218; [2002] QB 874[1]), the Court of Appeal was of the same opinion: see pp884-886. It is at this point that I respectfully differ from both the lower courts. Qualifying user having been found, there was nothing in the material before the council to support the conclusion that such user had been otherwise than as of right within the meaning of section 22 of the 1965 Act.

[9] The foregoing paragraphs of this opinion are directed to the issue that was contested before the lower courts and debated between the parties on the hearing of this appeal. After the House had reserved judgment at the conclusion of oral argument, however, the House became concerned to explore the possibility that, on the special facts of this case, the inhabitants of the locality might have indulged in lawful sports and pastimes for the qualifying period of 20 years or more not "as of right" but pursuant to a statutory right to do so. Such use would be inconsistent with use as of right. Counsel were invited to make written submissions on the point, which had not been raised or investigated below, and the House heard further oral argument on it. The House is grateful to counsel for responding so fully to its invitation, and consideration has been given to every statutory provision that appeared to be potentially relevant. In the event, I do not find it necessary to review those provisions in detail, since it is, to my mind, clear that none of them, on the facts found or agreed, can be relied upon to confer upon the local inhabitants a legal right to use the land for indulgence in lawful sports and pastimes. Indeed,

1 [2001] 3 PLR 120

Mr Philip Petchey, for the council, who had not himself sought to raise this contention earlier, found it hard to argue otherwise.

[10] For these reasons, and for those given by my noble and learned friends Lord Scott, Lord Rodger and Lord Walker, I would allow this appeal.

**LORD HUTTON**:

My lords,

[11] I have had the advantage of reading in draft the speech of my noble and learned friend, Lord Walker of Gestingthorpe, and for the reasons that he gives, and also for the reasons given by my noble and learned friends, Lord Bingham of Cornhill and Lord Rodger of Earlsferry, I, too, would allow this appeal.

**LORD SCOTT OF FOSCOTE**:

My lords,

*Introduction*

[12] The issue in this case is whether the use by the local inhabitants of a piece of land, commonly known as the sports arena, in Washington, Tyne and Wear, has turned that land into a "town or village green", as defined by section 22(1) of the Commons Registration Act 1965. Section 22(1) defines "town or village green" as including:

> land… on which the inhabitants of any locality have… indulged in [lawful] sports and pastimes as of right for not less than twenty years.

[13] Three years ago, your lordships had to consider the same issue. The case was *R* v *Oxfordshire County Council, ex parte Sunningwell Parish Council* [2000] 1 AC 335. It was the first time that your lordships had had to consider the section 22(1) definition. The present case is the second time.

[14] The main issue in *Sunningwell* was whether the inhabitants, whose use of the land for sports and pastimes was relied upon as constituting the requisite use "as of right", had to use the land in the belief that they had the right to do so. The House held that they did not have to have a personal belief in their right to use the land. It was sufficient that their use of the land, objectively evaluated, appeared to be a use as of right. The issue that arises in the present case is different. The issue is whether a use that is tolerated, and, indeed, encouraged, by the landowners, can none the less be a use as of right for the purposes of section 22(1). The issue is complicated in the present case by the circumstance that the successive owners of the sports arena during the period over which the use relied upon has taken place have been public authorities, holding the land for public purposes, and whose tenure of the land has been subject to various statutory provisions whose relevance and effect I must later consider. None the less, the core issue is whether the use relied upon has been use as of right.

[15] The leading opinion in *Sunningwell* was given by my noble and learned friend, Lord Hoffmann. Each of the other members of the Appellate Committee agreed with his opinion. It contains a valuable and

scholarly exposition of the historical provenance of the expression "as of right" in the 1965 Act that is as pertinent to this case as to *Sunningwell*. I cannot improve upon, and need not repeat, what Lord Hoffmann has said: see pp349-355.

[16] It is accepted, *per* Scott LJ in *Jones* v *Bates* [1938] 2 All ER 237, at p245, cited by Lord Hoffmann at p355D, that:

> [the words "as of right"] import the absence of any of the three characteristics of compulsion, secrecy or licence "nec vi, nec clam, nec precario," phraseology borrowed from the law of easements...

The issue in the present case is whether the use by the inhabitants was *nec precario*. Was an implied permission given by the landlord? If so, is use pursuant to an implied permission fatal to the contention that the inhabitants' use was as of right? How, if at all, does the fact that the sports arena was, throughout the period of use, public land held by public authorities for public purposes bear upon the answer to the question of whether the use was as of right? These questions raise some difficult issues. But let me start with the facts.

*Facts*

[17] The sports arena is a grass arena of 10 acres or thereabouts. It was acquired by Washington Development Corporation (WDC) in the course of their development of Washington new town pursuant to the New Towns Act 1965. WDC's Washington New Town Plan 1973 identified the land as "parkland/open space/playing field". In 1974, WDC, using excavated soil from the development of a shopping centre, laid out and grassed over the area. It would thereby have become recognisable as what is now the sports arena. It has never been fenced and it seems likely that public use of it for the purpose of recreation began shortly after the grassing over. In this litigation, however, the public recreational use contended for, and established by the evidence, is a use as from 1977.

[18] In 1977, WDC installed a double row of wooden benches, sufficient to accommodate 1,100 people, around the north, west and south perimeters of the sports arena. This was done in order to provide seating for the public on the occasion of a royal visit. A non-turf cricket wicket was laid down in 1979, and, over the years, the sports arena has been used for various recreational activities, ranging from team games to the walking of dogs.

[19] Title to the sports arena was, in 1989, transferred by WDC to the Commission for the New Towns (CNT) and, in 1996, was transferred by the CNT to Sunderland City Council. Throughout the period since the sports arena was grassed over in 1974, the owners for the time being, first WDC, then the CNT and, since 1996, the council, have mowed the grass during the summer.

*Litigation*

[20] On 24 December 1998, the council granted planning permission for the erection of a college of further education on land that includes the sports arena. It is common ground that the council want to dispose

of the land for use for that purpose. This proposal has been opposed by a number of local residents who have been accustomed to use the sports arena for recreational activities and who want to go on doing so. Their opposition to the grant of planning permission having failed, they made an application, on 18 November 1999, for the sports arena to be registered under the 1965 Act as a town or village green. The 1965 Act requires every "registration authority" to maintain a register of town or village greens: see section 3(1)(*b*). The registration authority for the area in which the sports arena is situate is the council: see section 2(1)(*a*). Section 13 of the 1965 Act enables the register to be amended where any land becomes a town or village green. The applicants' contention was that the sports arena had become a town or village green as a result of the requisite use of it by local inhabitants for at least 20 years. The council refused the application. They did so on the ground that the local inhabitants' use of the sports arena for recreational purposes had not been as of right but pursuant to an implied permission given by the landowners. Therefore, it was said, the use was not *nec precario*.

[21] An application was made by the appellant for judicial review of the council's refusal of the registration application. Smith J at [2001] 1 WLR 1327 refused the application. She held that the use had been pursuant to an implied permission and that that was sufficient, on the facts of the case, to disqualify the use from being as of right. She took into account, at p1340F, that the land was publicly owned:

> In my judgment, the fact that the land is in public ownership is plainly a relevant matter when one is considering what conclusion a reasonable person would draw from the circumstances of user. It is well known that local authorities do, as part of their normal functions, provide facilities for the use of the public and maintain them also at public expense. It is not part of the normal function of a private landowner to provide facilities for the public on the land. Public ownership of the land is plainly a relevant consideration.

I respectfully agree with those comments.

[22] The Court of Appeal ([2001] EWCA Civ 1218; [2002] QB 874) dismissed the appeal. Dyson LJ, with whose judgment the other two members of the court agreed, held, first, that, as a matter of principle, a claim that land had been used as of right could be defeated by showing that the use had been pursuant to an implied permission, and, second, that the council's conclusion, that there had been an implied permission, was a conclusion that the council, on the facts of the case, had been entitled to reach. On the point regarding the public ownership of the sports arena, Dyson LJ, although agreeing with Smith J that the public ownership was relevant, expressed the view, in [30] at p885E, that:

> on its own, it was a factor of little weight…

[23] On the further appeal to your lordships' House, Mr George Laurence QC, who did not appear below, concentrated upon attacking the proposition that use pursuant to an implied licence or permission could ever suffice to defeat a claim that the use was as of right. An express

licence or permission was, he said, essential. Mr Petchey, counsel for the respondent council, contended in answer that an implied licence would suffice to defeat an as of right claim and that the public use of the sports arena had been *precario*. Neither counsel dealt with the implications of the public ownership of the sports arena. This point emerged later, and the appeal was therefore restored for further written and oral submissions on the point.

*Statutory provisions relating to public authority land used for the purpose of public recreation*

[24] The New Towns Act 1981 (a consolidating Act) sets out the functions and powers of development corporations such as WDC and CNT. Section 21(1) applies to:

> Any land being, or forming part of, a common, open space or fuel or field garden allotment, which has been acquired for the purposes of this Act by a development corporation…

"Open space", as defined by section 80(1) of the 1981 Act, includes "any land… used for purposes of public recreation". Under subpara (a) of section 21(1), land to which section 21(1) applies may be used by the development corporation "or by any other person, in any manner in accordance with planning permission". This provision demonstrates the breadth of freedom that development corporations were intended to have in using or dealing with land that they had acquired for their statutory purposes. Not only were they themselves free to use the land "in any manner in accordance with planning permission", but so, too, were any persons to whom they might transfer the land: NB "or by any other person".

[25] Part II of the 1981 Act provides for the eventual dissolution of a development corporation and the vesting of its property in CNT: see section 41. The function of CNT is to "hold, manage and turn to account" the property of development corporations transferred to them under the Act: see section 36(1). CNT must have regard, *inter alia*, to the "convenience and welfare of persons residing, working and carrying on business" in the new town: see section 36(2)(a).

[26] These provisions seem to me to give rise to a number of issues on the facts of the present case. Does section 21 apply to land that was not, when acquired by the development corporation, being used for public recreation but where use for that purpose commenced after its acquisition? Mr Petchey expressly disclaimed, in answer to a question from me, any reliance upon section 21(1). In view of that disclaimer, your lordships cannot decide the point on this appeal. But, with respect to counsel, I do not think the answer to the point is plain. The sports arena was, at the date upon which WDC transferred it to CNT and at the date upon which CNT transferred it to the council, land "used for purposes of public recreation", that is, an "open space" as defined. The land had been acquired by WDC for the purposes of the Act (or its statutory predecessors). So why does section 21(1)(a) not apply and entitle the council to use the land "in any manner in accordance

with planning permission"? This question your lordships must leave unanswered.

[27] Sections 122 and 123 of the Local Government Act 1972 relate to land that has been acquired by a "principal council". The respondent council are a principal council: see section 270(1). Section 122(2A) (added by amendment under the Local Government, Planning and Land Act 1980) deals with the power of a principal council to appropriate land of various descriptions, including "open space" land, to other uses. Section 123(2A) (also added by amendment under the 1980 Act) deals with the power of a principal council to dispose of "open space" land. "Open space" is given the same definition as appears in the 1981 Act, and includes land "used for the purposes of public recreation": see section 270(1) of the 1972 Act and section 336(1) of the Town and Country Planning Act 1990. The two sections, 122 and 123, prescribe, however, special procedures that a council must follow if the "open space" land is to be appropriated to some other purpose or disposed of (as the case may be). The procedures include advertising the council's intention, allowing time for objections from members of the public and the giving of due consideration to any objections.

[28] It was, as I understood it, suggested by Mr Laurence that if the open space land had achieved the status of a 1965 Act town or village green, then, notwithstanding the disposal of the open space land by a principal council, the section 123(2A) procedures having been duly complied with, the land would retain its status as a town or village green under the 1965 Act. Mr Petchey did not contend that this was wrong. Your lordships do not need to decide the issue on this appeal, but, speaking for myself, I regard the proposition as highly dubious. An appropriation to other purposes duly carried out pursuant to section 122 would plainly override any public rights of use of an open space that had previously existed. Otherwise, the appropriation would be ineffective and the statutory power frustrated. The comparable procedures for a disposal prescribed by section 123 must surely bring about the same overriding effect.

[29] Finally, I should refer to section 10 of the Open Spaces Act 1906. Section 10 provides that:

> A local authority who have acquired any estate or interest in or control over any open space... under this Act shall, subject to any conditions under which the estate, interest, or control was so acquired –
>
> (*a*) hold and administer the open space... in trust to allow, and with a view to, the enjoyment thereof by the public as an open space within the meaning of this Act and under proper control and regulation and for no other purpose; and
>
> (*b*) maintain and keep the open space... in a good and decent state...
>
> "open space", as defined in section 20, includes '... land... which... is used for purposes of recreation...'

Section 123(2B)(*b*) of the Local Government Act 1972 enables open space land held under a 1906 Act trust to be disposed of freed from that trust.

[30] It is, I think, accepted that if the respondent council acquired the sports arena under the 1906 Act, the local inhabitants' use of the land for recreation would have been a use under the trust imposed by section 10 of the Act. The use would have been subject to regulation by the council, and would not have been a use as of right for the purposes of class c of section 22(1) of the Commons Registration Act 1965. But Mr Petchey accepted that Mr Laurence was correct in contending that the sports arena had not been acquired under the [1906] Act and that section 10 did not therefore apply. Here, too, although your lordships cannot, in view of this concession, conclude that Mr Laurence's contention is wrong, I do not, for myself, regard the point as clear. Is it necessary in order for open space land to have been acquired under the 1906 Act for it to be expressly so stated, whether in the deed of transfer or in some council minute?: *Attorney-General* v *Poole Corporation* [1938] Ch 23 is interesting on this point. The open space land in question had been conveyed to Poole Corporation:

> in fee simple to the intent that the same may for ever hereafter be preserved and used as a pleasure or recreation ground for the public use.

There was no express reference in the conveyance to the 1906 Act, but the Court of Appeal thought it plain that the Act applied. Indeed, counsel on both sides argued the case on the footing that that was so: see Sir Wilfrid Greene MR, at p30. It seems to me, therefore, that the 1906 Act should not have been set to one side in the present case simply on the ground that in the documents relating to the transfer to the council no express reference to the 1906 Act can be found. It would be, in my view, an arguable proposition that if the current use of land acquired by a local authority were use for the purposes of recreation, and if the land had not been purchased for some other inconsistent use and the local authority had had the intention that the land should continue to be used for the purposes of recreation, the provisions of section 10 would apply: cf counsel's argument in *Poole Corporation*, at p27. But your lordships cannot take the argument to a conclusion in the present case.

[31] The various statutory provisions to which I have referred are, in my opinion, whatever other relevance they may have, relevant as background against which the implications of the recreational use of the sports arena made by the local inhabitants from 1977 to, say, 1999 ought to be assessed. The sports arena, throughout that period, was, and remains, land in public ownership, held for public purposes, maintained at public expense and used by the public for recreation.

*Was the use "as of right" for section 22(1)(c) purposes?*

[32] It is accepted that: (i) the sports arena has been used for "lawful sports and pastimes"; (ii) the level of use has been sufficiently regular to satisfy section 22(1); (iii) the use has been made predominantly by inhabitants of the locality; and (iv) this use has continued for more than 20 years. What is in issue is whether the use has been "as of right". To that I must now turn. Before I do so, however, I would like to pay tribute to the council's director of administration, who prepared an admirably clear

report dated 19 April 2000 for the benefit of the special meeting of the council convened to deal with the registration application. I have drawn heavily on the report in describing the history of the sports arena and reciting the other facts relevant to this appeal.

[33] As Lord Hoffmann noted in *Sunningwell*, the concept of use as of right – *nec vi, nec clam, nec precario* – is derived from the law relating to the acquisition by prescription of private easements. Section 2 of the Prescription Act 1832 refers to rights of way or other easements "actually enjoyed by any person claiming right thereto without interruption for the full period of 20 years...". The concept was imported into the law relating to the dedication of land as a public highway. Section 1(1) of the Highways Act 1932 provided that "where a way... upon or over any land has been actually enjoyed by the public as of right and without interruption for a full period of 20 years, such way shall be deemed to have been dedicated as a highway unless there is sufficient evidence that there was no intention during that period to dedicate such way...": see, now, section 31(1) of the Highways Act 1980, which is in the same terms.

[34] It is a natural inclination to assume that these expressions, "claiming right thereto" (the 1832 Act), "as of right" (the 1932 Act and the 1980 Act), and "as of right" in the 1965 Act, all of which import the three characteristics – *nec vi, nec clam, nec precario* – ought to be given the same meaning and effect. The inclination should not, however, be taken too far. There are important differences between private easements over land and public rights over land and between the ways in which a public right of way can come into existence and the ways in which a town or village green can come into existence. To apply principles applicable to one type of right to another type of right without taking account of their differences is dangerous.

[35] If a private right of way is to be acquired by prescription, by 20 years' enjoyment by someone "claiming right thereto", use pursuant to a licence or permission from the owner of the land will usually – not invariably, but usually – be use that does not satisfy the *nec precario* condition.

[36] The acquisition of a private easement is the acquisition of a right *in rem* over land. If such a right is to be granted by a landowner, it must be granted by deed and the grant will usually be express. An easement can be acquired by implied grant only if the implication can be derived from the contents of a deed. A conveyance of land, for example, may carry with it the implied grant of easements necessary for the enjoyment of the land. But the conveyance will have been by deed and, accordingly, capable of effecting the grant of an easement. A mere agreement for the grant of an easement cannot by itself grant the easement.

[37] Where private easements are concerned, there are two exceptions to the requirement that the right must be granted by a deed. First, if permission to enjoy a right capable of constituting an easement is given by the landowner in terms likely to lead, and that do lead, the beneficiary of the permission to believe that he is entitled, on a permanent basis, to

enjoy the right and in that belief he sufficiently alters his position to his detriment, by expenditure of money or otherwise, he may become entitled in equity to the easement by proprietary estoppel: see *ER Ives Investment Ltd* v *High* [1967] 2 QB 379. The landowner would not be able to withdraw the permission he had given. Twenty years' enjoyment of the equitable right would surely enable the beneficiary of the permission to claim a legal easement under the 1832 Act. In such a case, it is easy to regard the enjoyment of the right pursuant to the original permission as enjoyment by a person "claiming right thereto". In such a case, the original permission would be the foundation of the claim of right but the enjoyment would not have been *precario*.

[38] Second, if an agreement to grant an easement were entered into for good consideration and the consideration were fully paid, the purchaser of the easement would at once become absolutely entitled in equity to the easement and would become entitled at law after 20 years' use. His enjoyment of the easement, although deriving from permission, would not have been *precario*, and, in my opinion, would have been enjoyment by a person "claiming right thereto": cf *Bridges* v *Mees* [1957] Ch 475, at pp484-485. It follows that the proposition that use pursuant to permission given by the landowner is always *precario* and cannot ever be as of right for prescription purposes is not correct.

[39] The same is true of use of a public way, or a would-be public way, following upon permission given by the landowner. A public right of way is not created by grant: it is created by dedication. The dedication does not have to be by deed and need not even be in writing. It can be evidenced by conduct. An implied permission for the public to use a particular path or track might be no more than a temporary, terminable permission, but equally it might indicate an intention to dedicate. An implied permission that sufficiently evidences an intention to dedicate creates the public right of way immediately. Twenty years' use by the public is not necessary. But 20 years' use "as of right" following a permission by a landowner that is indicative of an intention to dedicate will produce a deemed intention to dedicate unless the landowner can produce sufficient evidence that he had had no such intention: see section 1(1) of the 1932 Act and section 31(1) of the 1980 Act.

[40] There are differences, too, between public rights of way, on the one hand, and town or village greens, on the other. Public rights of way are created by dedication, express or implied or deemed. Town or village greens, on the other hand, must owe their existence to one or other of the three origins specified in section 22(1) of the 1965 Act. One of these is the 20 years' use as of right, to which I have already referred. Alternatively, a town or village green may be:

> land which has been allotted by or under any Act for the exercise or recreation of the inhabitants of any locality…

or

> land… on which the inhabitants of any locality have a customary right to indulge in lawful sports and pastimes…

In short, the origin of a town or village green must be either by statute or custom or 20 years' use. Dedication by the landowner is not a means by which a town or village green, as defined, can be created. So acts of an apparently dedicatory character are likely to have a quite different effect in relation to an alleged public right of way than in relation to an alleged town or village green. A

[41] The present case is concerned with implied permission. The installation and maintenance of the double rows of wooden benches around three sides of the sports arena and the regular cutting of the grass by the owners of the sports arena evidenced a clear enough willingness that the public should resort to the sports arena for recreational purposes. Indeed, it can reasonably be said that these acts encouraged the public to do so. Mr Petchey has submitted that since the public resorted to the sports arena pursuant to an implied permission from the landowners, its use of it during the 20-year period failed the *nec precario* requirement and was not as of right. B C

[42] Mr Laurence submitted that although use pursuant to an express permission would negate use as of right, use pursuant to a permission that was merely to be implied would not do so. Implied permission, he submitted, was to be equated with mere acquiescence or toleration on the part of the landowner. None of these, he submitted, would disqualify the use from being use as of right. Only an express permission would render the use *precario*. D

[43] My lords, I believe this rigid distinction between express permission and implied permission to be unacceptable. It is clear enough that merely standing by, with knowledge of the use, and doing nothing about it, that is, toleration or acquiescence, is consistent with the use being as of right. That that is so is accepted by Mr Petchey. But I am unable to accept either that an implied permission is necessarily in the same state as mere acquiescence or toleration, or that an implied permission is necessarily inconsistent with the use being as of right. Indeed, I do not, for the reasons that I have given, accept that even an express permission is necessarily inconsistent with use as of right. E F

[44] Lord Hoffmann, in *Sunningwell*, made it clear that the section 22(1) requirement of 20 years' use as of right did not require the users of the land to give evidence of their personal belief in their right of use. He said, at p356B: G

> A person who believes he has a right to use a footpath will use it in the way in which a person having such a right would use it. But user which is apparently as of right cannot be discounted merely because, as will often be the case, many of the users over a long period were subjectively indifferent as to whether a right existed, or even had private knowledge that it did not.

It is sufficient, therefore, if the use is "apparently as of right". But, of course, if the users do have a personal belief in their right to use the land, so much the better. H

[45] Permission for the public to use land for recreational purposes, or to pass along a path or track, may, depending upon the terms of the permission, if it is express, and upon the surrounding circumstances,

A whether or not it is express, indicate to the public that the permission is
temporary only, may be withdrawn, and is therefore precatory, or may
indicate to the public that its right of use is intended to be permanent.
In the case of a path or track, a sufficient indication, express or implied,
that the right of the public to use the path or track was intended to be
permanent, would usually constitute a dedication and create a public right
B of way. The members of the public using the way would be unlikely, not
having perused the *Halsbury* volume dealing with public highways, to
know anything about dedication or the manner in which public rights of
way can come into existence. They would simply use the way, following
the indications that they could do so, or following the example of others
who were using the way. Their use would at least be "apparently as
C of right". Their actual state of mind would not matter. The dedicatory
nature of the permission that the public could use the path or track would
positively support the contention that their user was as of right rather than
contradict it.

[46] Where a town or village green is concerned, however, a sufficient
indication, express or implied, that the right of the public to use the land
D for recreational purposes was intended to be permanent could not itself
endow the land with that status. But the quality of the use of the land by the
public, following the dedicatory indications in question, would surely be
as of right. It seems to me to be quite unreal to draw a distinction between
the quality of use of a path or track by members of the public following
an express or implied dedication and the quality of the recreational use by
E members of the public of a piece of land following permission given by a
landowner that, if dedication of land as a town or village green had been
possible, would have constituted a dedication. In each case, the quality
of the use, entirely consistent with the nature of the permission that had
been given, would have been "apparently as of right". The only difference
would have been that in the case of the public right of way the landowner
F could not, once the dedication had been accepted by public use, terminate
the use, but in the case of the land used for recreational purposes the
landowner could, provided that the 20 years had not expired, terminate
the use. But this difference does not seem to me to bear upon the quality
of the use of the land by the public in the meantime.

[47] Let me try to illustrate the point I am making by examples. If a
G landowner puts up a notice that says "The public may use this path as
a public highway", use by the public thereafter would surely be use as
of right. If a landowner puts up a notice that says "The public may use
this land for recreational purposes as a village green", use by the public
thereafter, until the landowner cancelled the notice and/or excluded the
public, would similarly be use as of right. Whether express or implied,
H permission to use a path over land or to use land for recreational
purposes may be of a sufficiently dedicatory character to justify the
same conclusion, namely that use by the public thereafter is use as
of right.

[48] I agree with Mr Petchey that, in the present case, the attitude of
the successive owners of the sports arena to the public use of the land for

recreation was more than mere acquiescence or toleration. There was, I A
agree, positive encouragement. The provision of the rows of benches was to make more comfortable the watching of the activities of others. The cutting of the grass was in order to enhance the enjoyment of the sports arena by those using it. I am receptive to the submission that the successive owners had impliedly consented to the recreational use of the land by the public. The users were, in my opinion, certainly not trespassers. But this B
does not, in my opinion, answer the question as to whether the use was as of right or *nec precario*.

[49] Was there any sign that the permission was intended to be temporary or revocable? There was none. The fact that the land was publicly owned seems to me highly material. Neither WDC nor CNT nor the council were, or are, private landowners. Their respective functions C
were and are functions to be discharged for the benefit of the public. The provision of benches for the public and the mowing of the grass were, in my opinion, not indicative of a precatory permission but of a public authority, mindful of their public responsibilities and function, desirous of providing recreational facilities to the inhabitants of the locality. In these circumstances, there seems to me to have been every reason for the D
inhabitants of the locality who used the sports arena to believe that they had the right to do so on a permanent basis.

[50] Accordingly, the nature of the implied permission from the landowners that the evidence shows to have been present was not, in my opinion, such as to prevent the use of the sports arena by the public from being use as of right. The positive encouragement to the public to enjoy E
the recreational facilities of the sports arena, constituted, in particular, by the provision of the benches, seems to me not to undermine but rather to reinforce the impression of members of the public that their use was as of right.

[51] Smith J and the Court of Appeal were, in my respectful opinion, led astray by according the concept of permission and, thus, of implied F
permission, a rigidity of character and effect that is not justified. They concluded that because use pursuant to permission will sometimes or often or usually be inconsistent with use as of right, it will always be inconsistent with use as of right. The conclusion, my lords, must, in my opinion, depend upon the nature of the permission, objectively assessed or construed. To conclude that use pursuant to implied permission is G
inconsistent with use as of right may, in most cases, be correct. But the conclusion is an evidentiary one; it is not a rule of law. And, in the present case, it is not, in my opinion, a correct evidentiary conclusion.

[52] For these reasons, I would, on the basis upon which the case has been argued before your lordships, allow the appeal. I am, however, for reasons that will have appeared, uneasy about this conclusion. Where H
"open space" land comes into the ownership of a "principal council", I think there are strong arguments for contending that the statutory scheme under the Local Government Act 1972, whether or not the Open Spaces Act 1906 or section 21(1) of the New Towns Act 1981 are applicable, excludes the operation of section 22(1) of the Commons Registration Act

1965. But these arguments have not been addressed to your lordships. I think also, as at present advised, that the power of disposal of open space land given to principal councils by section 123 of the 1972 Act will trump any town or village green status of the land whether or not it is registered. But this, too, if the council wish to take the point, must be decided on another occasion.

**LORD RODGER OF EARLSFERRY**:

My lords,

[53] The town of Washington lies within the jurisdiction of Sunderland City Council (the council). From at least 1977, members of the public have used an area near the town centre – referred to as "the sports arena" — for recreation. In truth, it is just an open, flat area of grass of some 13 acres that Washington Development Corporation had laid out in around 1974. In the Washington New Town Plan 1973, the land was identified as "parkland/open space/playing field". In 1977, around the time of the Queen's silver jubilee visit to the ground, the development corporation constructed wooden seats along much of the perimeter. A hard-surface cricket pitch was laid out in 1979. For the rest, the public bodies that have owned the land — most recently, the council — have done little except to keep the grass cut. Local people have used the ground in their different ways. Toddlers have played there, children of all ages have kicked a ball around or played cricket and other games, a Sunday league football team has used it for its matches. Many have simply treated it as a place to picnic, socialise, take their ease in the sunshine or walk the dog.

[54] In 1999, the appellant, Mrs Pamela Beresford, sought to register the area as a town or village green under the Commons Registration Act 1965. In terms of the relevant part of the definition in section 22(1), as it then stood, a town or village green means land upon which the inhabitants of any locality have indulged in lawful sports and pastimes "as of right" for not less than 20 years. Having considered a well-reasoned and objective submission by their director of administration, the council, acting as the registration authority, refused Mrs Beresford's application, but only on the ground that although the land had indeed been used for lawful sports and pastimes for more than 20 years, the use had not been as of right but by virtue of an implied licence from the owners.

[55] In *R* v *Oxfordshire County Council, ex parte Sunningwell Parish Council* [2000] 1 AC 335, Lord Hoffmann explained, in illuminating detail, why the words "as of right" are to be interpreted in the same way in section 22(1) of the 1965 Act as in section 5 of the Prescription Act 1832 (as amended) and section 1(1) of the Rights of Way Act 1932. Long before, in *Gardner* v *Hodgson's Kingston Brewery Co Ltd* [1903] AC 229, at pp238 and 239, both Lord Davey, impliedly, and Lord Lindley, expressly, had held that these words in the 1832 Act were intended to have the same meaning as the older expression *nec vi, nec clam, nec precario*. Lord Hoffmann adopted that interpretation and translated the phrase as "not by force, nor stealth, nor the licence of the owner": see p350H. So, if the inhabitants of any locality have engaged in lawful sports and pastimes *nec vi, nec clam, nec precario* for

at least 20 years, they have engaged in them as of right and the land can be A
registered as a town or village green in terms of the 1965 Act.

[56] It is not suggested that members of the public used the sports arena *vi* — by force: the owners did not try to stop them and so there was no question of them overcoming any resistance on the owners' part. Equally, the public was not enjoying itself *clam* — by stealth: on the
contrary, they used the land openly and the owners knew what was going B
on. The council concluded, however, that the local residents and others enjoying the land had been doing so *precario*, by virtue of the licence of the owners of the land. Admittedly, there was nothing to show that the owners had given any express permission or licence to the public. But the facts as a whole, and the cutting of the grass, and the construction of the
seating in particular, showed that the owners had actively encouraged the C
use of the area for recreation and so had impliedly granted a licence, or given permission, for it to be used in that way. Use of the land by virtue of this licence or permission could not constitute use as of right for purposes of section 22(1) of the 1965 Act. Smith J at [2001] 1 WLR 1327 dismissed Mrs Beresford's application for *certiorari* to quash the council's decision,
and the Court of Appeal dismissed her appeal: see [2001] EWCA Civ 1218; D
[2002] QB 874.

[57] In Roman law, "*precarium*" is the name given to a gratuitous grant of enjoyment of land or goods that is revocable at will. The arrangement is informal, and is based upon the grantor's goodwill, whether more or less enthusiastic. But however informal, the arrangement does involve
a positive act of granting the use of the property, as opposed to mere E
acquiescence in its use. The name suggests, and the digest texts indicate, that, in Roman law, the paradigm case is of a grant in response to a request. The arrangement lasts for only so long as the grantor allows, *tamdiu quamdiu is qui concessit patitur*: D43.26.1 pr, Ulpian 1 *institutionum*. The concept of *precarium* crops up in different areas of Roman law, but,
importantly, in connection with interdicts. The *praetor* protects someone F
from interference if he has taken possession of land, or has begun carrying out work *nec vi, nec clam, nec precario*.

[58] In *De legibus et consuetudinibus Angliae,* Bracton took over the noun *precarium* and its congeners from the vocabulary of Roman law and used them in a number of contexts, but always with reference to a
gratuitous grant that is revocable at any time at the grantor's pleasure. G
See, for instance, lib 2 ff52 and 52b. In lib 4f 221, Bracton discusses the acquisition of easements by use for some time *nec vi, nec clam, nec precario* – the last being, he says, the same as *de gratia,* of grace. Under reference to the second of these passages, in speaking of the use of a watercourse in *Burrows* v *Lang* [1901] 2 Ch 502, at p510, Farwell J asked "What is
precarious?" and answered his own question: "That which depends, not H
on right, but on the will of another person." Some years before, in *Sturges* v *Bridgman* [1879] 11 ChD 852, at p863, Thesiger LJ had indicated that if a man "temporarily licenses" his neighbour's enjoyment, that enjoyment is *precario* in terms of the civil law phrase *nec vi, nec clam, nec precario*. It is important to notice that, in this regard, English law distinguishes between

an owner who grants such a temporary licence or permission for an activity and an owner who merely acquiesces in it: see *Gale on Easements* (17th ed) 2002, para 4-83. Someone who acts with the mere acquiescence of the owner does so *nec precario*.

[59] The council were, accordingly, entitled to refuse Mrs Beresford's application for registration of the area as a town or village green only if those who used the sports arena did so by the revocable will of the owners of the land, that is to say, by virtue of a licence that the owners had granted in their favour and could have withdrawn at any time. The grant of such a licence to those using the ground must have comprised a positive act by the owners, as opposed to their mere acquiescence in the use being made of the land. Prudent landowners will often indicate expressly, by a notice in appropriate terms or in some other way, when they are licensing or permitting the public to use their land during their pleasure only. But I see no reason, in principle, why, in an appropriate case, the implied grant of such a revocable licence or permission could not be established by inference from the relevant circumstances.

[60] In the present case, the owners did not expressly license the use of the land by the public. The council rely upon two circumstances, however, as justifying the inference that those who used the sports arena did so *precario*, merely by licence from the owners of the land. The first is that the owners cut the grass. But that is at least equally explicable on the basis that the owners were concerned, as many owners would be, for the appearance of such a large and prominent area of open land in the heart of the town. Like charity, care of amenities begins at home. The second matter relied upon is the now rather dilapidated wooden seating along the perimeter. Whatever may have been its original purpose, the continued existence of the seating is consistent with the owners of the land having acquiesced, perhaps quite happily, in people using the area for football or other games that their friends or relatives would wish, or feel obliged, to watch. To an extent, the owners may thus have encouraged these activities. The mere fact that a landowner encourages an activity on his land does not indicate, however, that it takes place only by virtue of his revocable permission. In brief, neither cutting the grass nor constructing and leaving the seating in place justifies an inference that the owners of the sports arena positively granted a licence to local residents and others, who were then to be regarded as using the land by virtue of that licence, which the owners could withdraw at any time.

[61] In these circumstances, I would conclude that local people used the land *nec precario*.

[62] After the first hearing of the appeal, however, your lordships invited further written and oral submissions from counsel upon whether any of the statutes that might apply to local authority land had conferred on the local residents and others a right to use the sports arena, with the result that their use would be "of right", as opposed to being "as of right" in terms of section 22(1) of the 1965 Act. Having considered those submissions, for the reasons given by my noble and learned friend Lord

Walker of Gestingthorpe, I am satisfied that, on the agreed facts, neither the designation of the land as "open space" in the new town plan nor any of the statutes conferred any such right in this case.

[63] It follows that the local residents and others indulged in their sports and pastimes on the sports arena "as of right" in terms of section 22(1). Mrs Beresford is accordingly entitled to have the land registered under the 1965 Act as a town or village green.

[64] In a memorable passage in *Napier's Trustees* v *Morrison* (1851) 13 D 1404, at p1409, dealing with a public right of way, Lord Cockburn deprecated the citation in the Court of Session of authorities from England. He really wished, he said — taking a swipe at a future lord president, among others — that Scottish counsel and judges:

> could imitate the example set us by the counsel and the judges of that kingdom, who decide their causes by their own rules and customs, without exposing themselves by referring to foreign systems, the very language of which they do not comprehend.

Times change. In the course of the hearing of this appeal, well-informed counsel on both sides referred your lordships to a number of Scottish authorities on the acquisition of servitudes and public rights of way. In *Mann* v *Brodie* (1885) 10 App Cas 378, at pp385-387, Lord Blackburn analysed some of the differences between the English and Scottish law on the topic. Lord Hoffmann referred to that discussion in *R* v *Oxfordshire County Council, ex parte Sunningwell Parish Council* [2000] 1 AC 335, at p352. While exercising all due caution, and at the risk of disturbing the shade of Lord Cockburn, I believe that the Scottish authorities can provide some assistance in this case, at least by way of confirming the conclusion that I have already reached.

[65] The phrase "*nec vi, nec clam, nec precario*", taken over from Roman law, has resounded just as powerfully among Scottish lawyers and judges as among their brethren south of the border. But in reading the Scottish cases, a linguistic point must be noted. English judges have tended to use "tolerance" as a synonym for "acquiescence": see, for instance, *Mills* v *Silver* [1991] Ch 271. Scottish judges, on the other hand, have tended to use the term "tolerance" as a synonym for "permission" and as a translation of "*precarium*". This is perfectly understandable, since an owner who, perhaps somewhat reluctantly, decides to permit the public to walk across his land until further notice may be said to "tolerate" them doing so. That is what Lord Cockburn had in mind when he said, in *Napier's Trustees*, at p1408, that the defenders have possessed the road "by no trespass or tolerance, but as a public road". Similarly, in a different context, in *Scottish Property Investment Co Building Society* v *Horne* (1881) 8 R 737, at p740, Lord President Inglis said that to warrant the remedy of summary ejection, the defender's possession of premises had to be vicious, that is, obtained by fraud or force, or precarious possession. He added: "A precarious possession is a possession by tolerance merely." It was in that sense that Lord Kinnear, a recognised authority on Scottish land law, used the phrase "tolerance or permission" in *Folkestone Corporation* v *Brockman* [1914] AC 338, at p356.

[66] In *Marquis of Bute* v *McKirdy & McMillan Ltd* 1937 SLT 93, for some 70 years the public on the Isle of Bute had used a track to pass from a public road to part of the foreshore for purposes of bathing and recreation. The Marquis of Bute, who owned the relevant land, contended that the use of the track by the public should be attributed to the tolerance of himself and his predecessors in title. He therefore sought an interdict against a bus company that had been bringing large numbers of trippers to the point on the public road from which they could use the track to get to the beach. Rejecting the pursuer's contention, Lord President Normand held, at pp119-120, that the proper question was whether:

> having regard to the sparseness or density of the population, the user over the prescriptive period was in degree and quality such as might have been expected if the road had been an undisputed right of way. If the public user is of that degree and quality, the proprietor, who fails for the prescriptive period to assert or to put on record his right to exclude the public, must be taken to have remained inactive, not from tolerance, but because the public right could not have been successfully disputed or because he acquiesced in it.

The First Division of the Court of Session, having concluded that the bus company had proved the existence of a public right of way for pedestrians, pronounced decree of absolvitor in their favour.

[67] In *Cumbernauld and Kilsyth District Council* v *Dollar Land (Cumbernauld) Ltd* 1992 SLT 1035, the council raised an action of declarator that a public right of way existed over a raised walkway crossing the centre of Cumbernauld. The walkway, which the defenders had bought, along with other properties, from Cumbernauld Development Corporation, was extensively used by the public to get from one part of the town to another. Holding that a public right of way had been established, Lord President Hope observed, at p1042I:

> the occasional or irregular use of a path by hillwalkers or by others who resort to the countryside can readily be distinguished from the continuous use of it by members of the public as a route from one public place to another. It seems to me to be clear, on an examination of all the later authorities, that a proprietor who allows a way over his land to be used by the public in the way the public would be expected to use it if there was a public right of way cannot claim that that use must be ascribed to tolerance, if he did nothing to limit or regulate that use at any time during the prescriptive period.

In dismissing the appeal to this House from the decision of the First Division (1993 SC (HL) 44, at p47A-D), Lord Jauncey adopted and approved both this passage from the opinion of Lord President Hope and the passage that I have quoted from the opinion of Lord President Normand in *Marquis of Bute*. Lord Jauncey went on to note, at pp47H-48A, that there is no principle of law that requires that there be conflict between the interest of the users of the right of way and those of a proprietor. If acquiescence could lead to a public right of way being established, "encouragement can even more readily be said to have the same consequences".

[68] Similarly, in the present case, for at least 20 years before Mrs Beresford made her application, the inhabitants of Washington had played and passed the time on the sports arena in the way in which they could have been expected to do as of right on a town or village green. Therefore, in the absence of any act on the owners' part to regulate the activities on the land or otherwise to show that the inhabitants were disporting themselves only by the owners' revocable leave or licence, it is proper to infer that the owners had acquiesced in the inhabitants' use of the land as of right. The same result follows if the owners are thought to have encouraged the activities.

[69] For these reasons, as well as those given by my noble and learned friends Lord Bingham of Cornhill and Lord Walker of Gestingthorpe, I would allow the appeal.

**LORD WALKER OF GESTINGTHORPE**:

My lords,

[70] The crucial issue in this appeal turns on the words "as of right" in the definition of "town or village green" in section 22(1) of the Commons Registration Act 1965. I set out the definition with the insertion of paragraph numbers that are not in the Act, but which are often used as a convenient means of denoting its three limbs:

(a) land that has been allotted by or under any Act for the exercise or recreation of the inhabitants of any locality; or

(b) upon which the inhabitants of any locality have a customary right to indulge in lawful sports and pastimes; or

(c) upon which the inhabitants of any locality have indulged in such sports and pastimes as of right for not less than 20 years.

[71] It might be supposed that there is, after the magisterial speech of Lord Hoffmann in *R* v *Oxfordshire County Council, ex parte Sunningwell Parish Council* [2000] 1 AC 335, little more to be said on the subject. Certainly, any consideration of the subject must start with Lord Hoffmann's speech, in which the rest of your lordships' House has concurred. But on the undisputed facts of this case (as to which, I gratefully adopt the summary in the speech of my noble and learned friend Lord Scott of Foscote) a new issue has been raised, that of implied licence (or permission or consent). That was the ground upon which Sunderland City Council succeeded before the judge ([2001] 1 WLR 1327) and in the Court of Appeal ([2001] EWCA Civ 1218; [2002] QB 874).

[72] It has often been pointed out that "as of right" does not mean "of right". It has sometimes been suggested that its meaning is closer to "as if of right": see, for instance, Lord Cowie in *Cumbernauld and Kilsyth District Council* v *Dollar Land (Cumbernauld) Ltd* 1992 SLT 1035, at p1043, approving counsel's formulation. This leads at once to the paradox that a trespasser (so long as he acts peaceably and openly) is in a position to acquire rights by prescription, whereas a licensee, who enters the land with the owner's permission, is unlikely to acquire such rights. Conversely, a landowner who puts up a notice stating "Private Land — Keep Out" is in a less strong position, if his notice is ignored by the public, than a landowner whose notice is in friendlier terms: "The public have permission to enter

this land on foot for recreation, but this permission may be withdrawn at any time."

[73] In *Gardner* v *Hodgson's Kingston Brewery Co Ltd* [1903] AC 229, at p231, the Earl of Halsbury LC referred to the phrase "as of right" used in section 5 (and reflected in section 2) of the Prescription Act 1832, and observed:

> I cannot help thinking there has been a certain play upon words in commenting upon them. In a certain sense a man has a right to enjoy what he has paid for, and, therefore, if the appellant here at any time during the year when she had paid for the right to use this way had been hindered, she would have had a right to complain that what I will call her contract had been broken, and that during the year she had a right to use the way. I do not think that this would have established a right in the proper sense, because, being but a parol licence, it might be withdrawn, and her action would be for damages, but she would have no *right* to the way. And in no sense could the right be the right contemplated by the Act. That right means a right to exercise the right claimed against the will of the person over whose property it is sought to be exercised. It does not and cannot mean a user enjoyed from time to time at the will and pleasure of the owner of the property over which the user is sought.

[74] In that case, the party claiming a right of way through the yard of a neighbouring inn, and her predecessors in title, had, for well over 40 years, used the inn yard (the only means of access with carts and horses to her premises) and had paid the annual sum of 15 shillings to the innkeeper. The most likely explanation of this payment was as an acknowledgement of the innkeeper's title, amounting (as it was put by Lord Lindley, at p239) to:

> a succession of yearly licences not, perhaps, expressed every year, but implied and assumed and paid for...

So to make a charge for entry to land is one way of making it clear that entry is not as of right. The paying entrant would be there by licence, even though he would (as Lord Halsbury pointed out) have the right to complain if the landowner broke the terms of his contract.

[75] An entry charge of this sort can aptly be described as carrying with it an implied licence. The entrant who pays and the man on the gate who takes his money both know what the position is without the latter having to speak any words of permission, although he might qualify the permission by saying that no dogs, or bicycles, or radios are allowed. Similarly (especially in a small village community where people know their neighbours' habits), permission to enter land may be given by a nod or a wave, or by leaving open a gate or even a front door. All these acts could be described as amounting to implied consent, although I would prefer (at the risk of pedantry) to describe them as the expression of consent by non-verbal means. In each instance, there is a communication by some overt act that is intended to be understood, and is understood, as permission to do something that would otherwise be an act of trespass.

[76] The authorities contain many references (which can be identified A and understood more readily since *Sunningwell*) to the importance of looking at the overt conduct of those involved, including what the landowner said and did from time to time during the period that the court has to examine. If the landowner found that his land was being used as a footpath by his neighbour (in a private right-of-way case) or by the whole village (in a public right-of-way case) and he suffered in silence, he would B be treated as having acquiesced in what was going on. As Fry J (one of the judges who advised the House of Lords in *Dalton* v *Henry Angus & Co* [1881] LR 6 App Cas 740) said, in that case, at p773:

> the whole law of prescription and the whole law which governs the presumption or inference of a grant or covenant rest upon acquiescence. The Courts and the Judges have had recourse to various expedients for quieting C the possession of persons in the exercise of rights which have not been resisted by the persons against whom they are exercised, but in all cases it appears to me that acquiescence and nothing else is the principle upon which these expedients rest...

(Lord Blackburn took a different view about acquiescence — see D pp817-818 — but the view expressed by Fry J seems to have prevailed.)

[77] A landowner who wishes to stop the acquisition of prescriptive rights over his land must not acquiesce and suffer in silence. The Lord President, Lord Hope, put the point clearly in the Inner House in *Cumbernauld*, at p1041G (that case was concerned with section 3 of the Prescription and Limitation (Scotland) Act 1973, which does not use the phrase "as of right", but it is common ground that there is still such a E requirement under the law of Scotland):

> where the user is of such amount and in such manner as would reasonably be regarded as being the assertion of a public right, the owner cannot stand by and ask that his inaction be ascribed to his good nature or to tolerance. If his position is to be that the user is by his leave and licence, he F must do something to make the public aware of that fact so that they know that the route is being used by them only with his permission and not as of right...

Lord Jauncey of Tullichettle quoted that passage with approval when the case came before your lordships House on a further appeal (1993 SC (HL) 44, at p47); the rest of the House concurred in the speech of Lord G Jauncey.

[78] Later in his judgment in the Inner House, Lord Hope said, at p1042J:

> a proprietor who allows a way over his land to be used by the public in the way the public would be expected to use it if there was a public right of way cannot claim that that use must be ascribed to tolerance, if he did nothing to H limit or regulate that use at any time during the prescriptive period...

Mr Laurence QC (for the appellants) emphasised Lord Hope's repeated references (in the two passages set out above, and again at p1042L) to the need for the landowner to *do* something.

[79] Acquiescence, by contrast, denotes passive inactivity. The law sometimes treats acquiescence as equivalent in its effect to actual consent. In particular, acquiescence may lead to a person losing his right to complain of something just as if he had agreed to it beforehand. In this area of the law, it would be quite wrong, in my opinion, to treat a landowner's silent passive acquiescence in persons using his land as having the same effect as permission communicated (whether in writing, by spoken words, or by overt and unequivocal conduct) to those persons. To do so would be to reward inactivity. Despite his failing to act, and, indeed, simply by his failure to act, the landowner would change the quality of the use being made of his land from use as of right to use that is (in the sense of the Latin maxim) precarious.

[80] This point was put very clearly, and, to my mind very compellingly, by Dillon LJ in *Mills* v *Silver* [1991] Ch 271, at pp279G-280. After referring to what the judge at first instance had said about tolerance, Dillon LJ observed:

> The topic of tolerance has bulked fairly large in recent decisions of this court dealing with claims to prescriptive rights, since the decision in *Alfred F Beckett Ltd v Lyons* [1967] Ch 449. If passages in successive judgments are taken on their own out of context and added together, it would be easy to say, as, with all respect, it seems to me that the judge did in the present case, that there is an established principle of law that no prescriptive right can be acquired if the user by the dominant owner of the servient tenement in the particular manner for the appropriate number of years has been tolerated without objection by the servient owner. But there cannot be any such principle of law because it is, with rights of way, fundamentally inconsistent with the whole notion of acquisition of rights by prescription. It is difficult to see how, if there is such a principle, there could ever be a prescriptive right of way. It follows that the various passages in the judgments in question cannot be taken on their own out of context. If each case is looked at on its own and regarded as a whole, none lays down any such far-reaching principle...

[81] Parker and Stocker LJJ both agreed with Dillon LJ, although each added some further reasons. Parker LJ referred to what Lord Halsbury had said in *Gardner* (in the passage that I have already quoted) and said, at p289, that by "against the will of the person" Lord Halsbury meant no more than "without the licence of the owner". Stocker LJ stated, at p293B:

> It seems clear from the passage in the judgment cited by Dillon LJ that the judge in the instant case failed to recognise the very limited circumstances in which the word "toleration" has been used in the cases cited which might be summarised as relating to the exercise of a purported right which was casual or trivial or in respect of which some form of consent for the user was established so that acquiescence did not arise...

I respectfully agree with both these observations. Stocker LJ was making the same point as Dillon LJ, that, in this context, consent is not a synonym for acquiescence, but almost its antithesis: the former negatives user as of right, whereas the latter is an essential ingredient of prescription by user as of right.

[82] Smith J referred to *Mills* although it had not been cited to her. It was cited in the Court of Appeal, but was not referred to by Dyson LJ. It was referred to with approval by Lord Hoffmann in *Sunningwell.* For my part, I have found it, after *Sunningwell,* the most helpful guide to the relevant principles.

[83] In the Court of Appeal, Dyson LJ considered that implied permission could defeat a claim to user as of right, as Smith J had held at first instance. I can agree with that as a general proposition, provided that the permission is implied by (or inferred from) overt conduct of the landowner, such as making a charge for admission or asserting his title by the occasional closure of the land to all-comers. Such actions have an impact upon members of the public and demonstrate that their access to the land, when they do have access, depends upon the landowner's permission. But I cannot agree that there was any evidence of overt acts (on the part of the city council or their predecessors) justifying the conclusion of an implied licence in this case.

[84] The grounds of the licensing committee's decision, based upon the report by the director of administration, were as follows:

> (a) Members were satisfied that evidence showed the use of the sports arena for "lawful sports and pastimes" by the inhabitants of Washington for a period of at least 20 years prior to the making of the application, the level of use being more than trivial or sporadic. The real issue for consideration was whether there had been permission or a licence to use the site in this way.
>
> (b) Having taken legal advice, members were satisfied that an implied licence would be sufficient to defeat the application, provided that there was sufficient evidence to support the existence of a licence.
>
> (c) Members considered that there was evidence of an implied licence since the site is publicly owned land, specifically laid out as an arena with seating, which is adjacent to the Princess Anne Park and which has been maintained by the council and the Washington Development Corporation before it. Members agreed with the comment in the report that "it is difficult to conceive that anyone could have imagined that this was other than a recreational area, provided for use by the public for recreation". The other information contained in section 2 of the report, whilst not in itself conclusive, supported the view that the Sports Arena was intended for public use...

[85] In my opinion, this reasoning, and the fuller reasoning in the director's report that it was based upon, must be regarded as erroneous. The fact that the city council and its predecessors were willing for the land to be used as an area for informal sports and games, and provided some minimal facilities (now decaying) in the form of benches and a single hard cricket pitch, cannot be regarded as overt acts communicating permission to enter. Nor could the regular cutting of the grass, which was a natural action for any responsible landowner. To treat these acts as amounting to an implied licence, permission or consent would involve a fiction, like the fiction under which the placing or maintaining on land of an "allurement" would be regarded as an implied licence that might lead to a straying child being treated as an "invitee" rather than as a trespasser for the purposes of occupiers' liability: see generally *Herrington* v *British Railways Board* [1972] AC 877, especially the speech of Lord Diplock, at pp932-936. For

the reasons given by Dillon LJ in *Mills*, to add the fiction of implied licence to the unavoidable fiction of presumed grant would reduce this part of the law to a state of incoherence.

[86] I would, however, add that I feel some sympathy for the view taken by the courts below. The city council, as a local authority, are, in relation to this land, in a different position from a private landowner, however benevolent, who happens to own the site of a traditional village green. The land is held by the city council, and was held by their predecessors, for public law purposes. A local resident who takes a walk in a park owned by a local authority might indignantly reject any suggestion that he was a trespasser unless he obtained the local authority's consent to enter. He might say that it was the community's park, and that the local authority, as its legal owners, were (in a loose sense) in the position of a trustee with a duty to let him in. (Indeed, that is how Finnemore J put the position in *Hall* v *Beckenham Corporation* [1949] 1 KB 716, at p728, which was concerned with a claim in nuisance against a local authority, the owners of a public park, in which members of the public flew noisy model aircraft.) So the notion of an implied statutory licence has its attractions.

[87] After that approach had been suggested, there was a further hearing of this appeal in order to consider the effect of various statutory provisions that were not referred to at the first hearing, including, in particular, section 10 of the Open Spaces Act 1906, sections 122 and 123 of the Local Government Act 1972 and section 19 of the Local Government (Miscellaneous Provisions) Act 1976. Where land is vested in a local authority on a statutory trust under section 10 of the Open Spaces Act 1906, inhabitants of the locality are beneficiaries of a statutory trust of a public nature, and it would be very difficult to regard those who use the park or other open space as trespassers (even if that expression is toned down to tolerated trespassers). The position would be the same if there were no statutory trust in the strict sense, but land had been appropriated for the purpose of public recreation.

[88] Those situations would raise difficult issues, but, in my opinion, they do not have to be decided by your lordships on this appeal, and would be better left for another occasion. The undisputed evidence does not establish, or give grounds for inferring, any statutory trust of the land or any appropriation of the land as recreational open space. Counsel for Sunderland rightly did not argue for some general implied exclusion of local authorities from the scope of section 22 of the Commons Registration Act 1965.

[89] It is worth summarising the salient points of the evidence.

(a) The land was first acquired by Washington Development Corporation (WDC) as part of what seems to have been an extensive acquisition under the very wide powers in the New Towns Act 1965. WDC did not acquire this particular area of land for any specific purpose, and were not under an obligation to appropriate it for any specific purpose, such as housing, public buildings, or open space. The plans for the new town provided for the area to be included in a sports complex consisting of an indoor leisure centre and an indoor swimming pool, both of which were built, and a tartan running track (enclosing a football field) and a grandstand, which were not built. Had the track and the grandstand been built, public access to them would no doubt have been regulated in some way, probably including charging an entrance fee. The area would have been devoted to recreation, but local inhabitants would not have used it as of right. A

(b) The ambitious plans for the sports complex have never been fully realised, but they still seem to have been regarded as at least a possibility in 1982 (when a manuscript draft report referred to an unencouraging opinion from the Sports Turf Research Institute) and in 1983, when the city council, although not yet owners of the land, referred to "the accommodation of a running track" in a report entitled *Open Space Recreation*. In the meantime, recreational use of the area by local inhabitants was tolerated (but not, for reasons that I have already stated, enjoyed by any overt licence). B C

(c) The land was transferred by WDC to the Commission for the New Towns (CNT) in 1989 as part of a general disposal of WDC's assets. It appears that CNT retained the land in 1991 (when other assets were transferred to the city council) because it was regarded as having potential for commercial development: see para 2.4 of the report by the director of administration. D

(d) When the land was eventually transferred by CNT to the city council in 1996, its use was restricted by covenant to "the provision of magistrates' courts and/or community health facilities and/or community leisure/recreation and/or other similar community related uses and developments": see para 2.7 of the same report. E

[90] In short, there is no evidence of any formal appropriation of the land as recreational open space by the city council or its predecessors. Nor is there material from which to infer an appropriation. Such action by WDC or CNT would have been unnecessary, and at or after the city council's acquisition in 1991 an appropriation as open space would have been inconsistent with the site's perceived development potential. It is true that the public's interim use of the land for recreation was not inimical to the city council's interests. But user can be as of right even though it is not adverse to the landowner's interests. F

[91] That was established by the decision of this House in *Cumbernauld*, at pp47F-48A, where Lord Jauncey said:

> senior counsel for the appellants argued that unless a public user of a way was adverse to the interests of the proprietor it must necessarily be ascribed to tolerance and that since the user of [a pedestrian walkway in the middle of a new town] had been positively encouraged by the development corporation, it could not amount to user as of right. For a user to be so considered there must, it was argued, be conflict between the interest of the users and that of the proprietor. For this somewhat stark proposition counsel could produce no authority. G H
>
> There is no principle of law which requires that there be conflict between the interest of users and those of a proprietor. As Lord President Normand pointed out in *Marquis of Bute v McKirdy and McMillan* acquiescence on the part of a proprietor in continued user throughout the prescriptive period without taking steps to assert or record his right of exclusion will result in

A the constitution of a public right of way against him. If acquiescence in these circumstances produces such a result encouragement can even more readily be said to have the same consequences…

[92] For these reasons, and for the further reasons set out in the speeches of my noble and learned friends Lord Bingham of Cornhill and Lord Rodger of Earlsferry, I would allow this appeal and quash the decision that B the city council took by their licensing committee. I reach this conclusion with mixed feelings. The campaigning group named Washington First may feel that they have won a famous victory, and saved an important public amenity from being built upon. That seems to be the likely consequence of this case. But the campaigners have achieved that end by a route that has bypassed normal development controls, and in a way that C might be thought to stretch the concept of a town or village green close to, or even beyond, the limits that parliament is likely to have intended. Any change in the law is of course a matter for parliament, but I respectfully agree with Lord Bingham's observations as to the need for care on the part of decision makers, whose conclusions as to the existence of a town or village green may have very important practical consequences. I also D respectfully agree with Lord Rodger's observations as to the assistance to be derived from the Scottish authorities, provided that note is taken of the different meanings in which "tolerance" has been used in England and Scotland respectively.

*Appeal allowed.*

# Gloucestershire County Council *v* Keyway (Gloucester) Ltd

[2003] EWHC 3012 (Admin)

QUEEN'S BENCH DIVISION: ADMINISTRATIVE COURT
*Sullivan J*
14 November 2003

**Stop notice – Service – Validity – Town and Country Planning Act 1990 – Defence under section 183(3) – Whether stop notice left at company's registered and principal office validly served – Whether notice must be brought to attention of responsible officer – Whether "person" in section 233(2) of Local Government Act 1972 meaning "natural" not "corporate" person**

The appellant council served a stop notice on the respondent, pursuant to section 183 of the Town and Country Planning Act 1990. It had been left at the reception desk of the respondent's registered and principal office with an unidentified person. Following non-compliance with the notice, the council preferred an information against the respondent alleging contravention. Before the justices, the respondent contended, *inter alia*, that service had not been effected. It relied upon the defence, pursuant to section 187(3) of the 1990 Act, that it did not know, and could not reasonably have been expected to have known, of the existence of the stop notice. It argued that, because a breach of a stop notice constituted a criminal offence, a restricted approach should be adopted when interpreting section 233 of the Local Government Act 1972; this section provides that "such a document *may* be given to or served on the person in question either by delivering it to him, or by leaving it at his proper address..." (emphasis supplied). Accordingly, it was not sufficient to serve a notice on a company by leaving it at its registered or principal office, where it could be overlooked; the section meant a natural not a corporate person, and the word "may" should be read as "must". On a preliminary issue, the justices concluded that the notice had not been served on the respondent, and dismissed the information without hearing the substantive merits of the case. The council appealed.

***Held*** The stop notice was served on the respondent under section 233 of the 1972 Act.

The answer to how a document may be served on a company secretary is found in section 233(2) of the 1972 Act. The methods include leaving it at the company secretary's proper address, which is defined in subsection (4)(*a*) as the company's "registered or principal office": see [25]. Service of a document by a local authority may lead to criminal liability, but that is no reason to adopt a strained and unnatural interpretation of section 233. Service is effected on corporate bodies by sending documents through the post. It is to be expected that companies will make the necessary

A administrative arrangements to ensure that the correct persons within the company hierarchy have sight of important documents: see [26]. It is not necessary for the purposes of effecting service under section 233 to prove that the document has actually been delivered into the hands of a director or secretary of the company. It is strictly unnecessary to consider the alternative method of service that is afforded by section 329 of the B 1990 Act.

**No cases were referred to in the judgment**

**Appeal against a decision of the justices for the county of Gloucester**

This was an appeal by Gloucestershire County Council by way of case stated against a decision of the Justices for the county of Gloucester, sitting C in Cheltenham on 10 March 2003, dismissing an information against Keyway (Gloucester) Ltd.

*Peter Wadsley (instructed by the solicitor to Gloucestershire County Council) appeared for the appellant council.*

*Ian Dove QC (instructed by Willans, of Cheltenham) represented the respondent,* D *Keyway (Gloucester) Ltd.*

*The following judgment was delivered.*

**SULLIVAN J:**

[1] This is a prosecutor's appeal by way of case stated against a decision by the Justices for the county of Gloucester sitting in Cheltenham on E 10 March 2003.

[2] On 28 November 2002, an information had been preferred by the appellants that:

> Keyway (Gloucester) Limited, did contravene a stop notice issued pursuant to section 183 of the Town and Country Planning Act 1990 in respect of land at Colemans Farm, Elmstone Hardwicke, Gloucestershire the stop notice F having been served on Keyway (Gloucester) Limited on 13th November 2002 which notice prohibited the importation and depositing of all waste material on land at Colemans Farm, Elmstone Hardwicke, Cheltenham, Gloucestershire [the site] contrary to section 187 of the Town and Country Planning Act 1990.

G [3] On 10 March 2003, the justices did not deal with the substantive merits of the information because they had been asked to deal with a preliminary issue: namely whether the stop notice had been served on the respondent company. They concluded that it had not and dismissed the information on that ground.

[4] The statutory scheme may be summarised as follows. Under section 172 of the Town and Country Planning Act 1990 (the Act), a local H planning authority may issue an enforcement notice where it appears to the authority that there has been a breach of planning control and that it is expedient to issue the notice.

[5] An enforcement notice must specify the date upon which it is to take effect (section 173(8)) and this must be at least 28 days after it has

been served (section 172(3)). If an appeal is made against the notice before it has taken effect, the notice will effectively be suspended until final determination of the appeal, which may take place many months later, particularly if there is an appeal to the High Court under section 289 of the Act.

[6] Since it may be desirable to bring unlawful development to an immediate halt, section 183 enables the local planning authority to serve a stop notice. Subsection (1) states:

> Where the local planning authority consider it expedient that any relevant activity should cease before the expiry of the period for compliance with an enforcement notice, they may, when they serve the copy of the enforcement notice or afterwards, serve a notice (in this Act referred to as a "stop notice") prohibiting the carrying out of that activity on the land to which the enforcement notice relates, or any part of that land specified in the stop notice.

[7] Section 184 provides:

> .–(1) A stop notice must refer to the enforcement notice to which it relates and have a copy of that notice annexed to it.
>
> (2) A stop notice must specify the date on which it will take effect (and it cannot be contravened until that date).
>
> (3) That date –
>
> (a) must be not earlier than three days after the date when the notice is served, unless the local planning authority consider that there are special reasons for specifying an earlier date and a statement of those reasons is served with the stop notice; and
>
> (b) must not be later than twenty-eight days from the date when the notice is first served on any person.
>
> ...
>
> (6) Where a stop notice has been served in respect of any land, the local planning authority may display there a notice (in this section and Section 187 referred to as a "site notice")...

Section 187(1) states:

> If any person contravenes a stop notice after a site notice has been displayed or the stop notice has been served on him he shall be guilty of an offence.

Subsection (3) makes provision for a defence:

> In proceedings for an offence under this section it shall be a defence for the accused to prove –
>
> (a) that the stop notice was not served on him, and
>
> (b) that he did not know, and could not reasonably have been expected to know, of its existence.

[8] Pausing there, it is not in dispute that an enforcement notice dated 13 November 2002 was served by the appellants on the respondent. The notice took effect (subject to any appeal by the respondent) on 13 December 2002, and required the respondent to stop importing and tipping waste materials on the site. The stop notice referred to in the

case stated was dated 13 November 2002, said that it took effect on 19 November and required the importation and depositing of waste on the site to cease on that date.

[9] Before turning to the appellants' endeavours to serve the stop notice on the respondent, it is helpful to set out the enactments that deal with service. Section 329 of the Act deals with "Service of Notices" and provides, so far as material, in subsection (1):

> Any notice or other document required or authorised to be served or given under this Act may be served or given either –
> (a) by delivering it to the person on whom it is to be served or to whom it is to be given; or
> (b) by leaving it at the usual or last known place of abode of that person, or, in a case where an address for service has been given by that person, at that address; or
> (c) by sending it in a prepaid registered letter, or by the recorded delivery service, addressed to that person at his usual or last known place of abode, or, in a case where an address for service has been given by that person, at that address; or
> ...
> (d) in the case of an incorporated company or body, by delivering it to the secretary or clerk of the company or body at their registered or principal office, or sending it in a prepaid registered letter, or by the recorded delivery service, addressed to the secretary or clerk of the company or body at that office.

[10] Subsection (4) provides:

> This section is without prejudice to section 233 of the Local Government Act 1972 (general provisions as to service of notices by local authorities).

[11] The Act was a consolidating Act, and the provisions of section 329(1)(d) re-enacted provisions that had been contained in the 1971, 1962 and 1947 Town and Country Planning Acts. First enacted in section 105(1)(*d*) of the 1947 Act, the only subsequent amendment was the addition of a reference to recorded delivery in section 214(1)(*d*) of the 1962 Act. Thereafter, the provision has remained unchanged: see section 283(1)(*d*) of the 1971 Act.

[12] Subsection (4) was added by the Planning and Compensation Act 1991. Section 233 of the Local Government Act 1972 provides, so far as is material:

> .–(1) Subject to subsection (8) below, subsections (2) to (5) below shall have effect in relation to any notice, order or other document required or authorised by or under any enactment to be given to or served on any person by or on behalf of a local authority or by an officer of a local authority.
> (2) Any such document may be given to or served on the person in question either by delivering it to him, or by leaving it at his proper address, or by sending it by post to him at that address.
> (3) Any such document may –
> (*a*) in the case of a body corporate, be given to or served on the secretary or clerk of that body;
> ...

> (4) For the purposes of this section and of section 26 of the Interpretation Act 1889 (service of documents by post) in its application to this section, the proper address of any person to or on whom a document is to be given or served shall be his last known address, except that –
>
> (*a*) in the case of a body corporate or their secretary or clerk, it shall be the address of the registered or principal office of that body…

Subsection (10) provides:

> Except as aforesaid and subject to any provision of any enactment or instrument excluding the foregoing provisions of this section, the methods of giving or serving documents which are available under those provisions are in addition to the methods which are available under any other enactment or any instrument made under any enactment.

[13] Pausing there, it is plain from subsection 329(4) of the Act and subsection 233(10) of the 1972 Act that the newer methods of service contained in the 1972 Act are additional to the older methods of service contained in section 329 of the Act. Thus, it was sufficient for the appellants to have effected service in accordance with either section 233 of the 1972 Act or section 329 of the Act. Mr Ian Dove QC, on behalf of the respondent, did not dissent from that proposition.

[14] In para 2 of the statement of case, the justices found that:

> On 13th November 2002 the Respondents' registered office was "Newbliss", Rudford, Gloucester and the respondents' principal office was Corinium House, Luther Challis Business Centre, Corinium Avenue, Barnwood, Gloucester. The Respondents' Company Secretary was Ms Sarah Catherine Christie.

The justices then set out their findings in respect of the registered office and the principal office. As to the former, they found:

> (c) Mr Worrallo (Process Server employed by the County Council) visited Newbliss Farm, Rudford, Gloucester on 13th November 2002, with an envelope containing a "stop notice" for service on Sarah Catherine Christie (Company Secretary). The envelope was correctly addressed.
>
> (d) On 13th November 2002 Mr Worrallo spoke to an unidentified woman at Newbliss Farm. This woman agreed to deliver the documents to Ms Christie (Company Secretary). Mr Worrallo described this lady to the court but did not obtain a name, signature or receipt for the document.
>
> (e) Service of this "stop notice" was not sent by registered or recorded delivery.
>
> (f) The "stop notice" was delivered by Mr Worrallo, (Process Server) by hand to an unidentified female at Newbliss Farm.
>
> (g) The unidentified female as described by Mr Worrallo (Process Server) did not fit the description of Ms Christie (Company Secretary).

[15] The justices' findings in respect of the principal office were as follows:

> (h) On 13th November 2002 the process server (Mr Worrallo) went to the principal office (Corinium House) with the "stop notice" contained in an envelope, which was correctly addressed.

> (i) The Process Server (Mr Worrallo) went through the front door of Corinium House and handed the envelope containing the "stop notice" to an unidentified male person standing behind the desk in the reception area.
>
> (j) Mr Worrallo (Process Server) did not obtain the name of this unidentified male person behind the reception desk at Corinium House, nor did he obtain a receipt.
>
> (k) Mr Worrallo (Process Server) left the envelope containing the "stop notice" with the unidentified man behind the reception desk with instructions to pass the envelope to a director or officer of Keyway (Gloucester) Limited. The Process Server (Mr Worrallo) left without obtaining a response whether in the negative or affirmative from the unidentified male.
>
> (l) The envelope containing the "stop notice" when left at the principal office by Mr Worrallo (Process Server) was not addressed specifically to Ms Christie (Company Secretary).
>
> (m) The reception desk at Corinium House is only manned by a female receptionist from 9.00 am till 5.30 pm, and the identity/position of the male within the company or even whether he was employed by the company was not clarified or confirmed by any witness.
>
> (n) Service of the "stop notice" was not sent by registered or recorded delivery.

[16] The justices then summarised the evidence given by Mr Worrallo, Ms Christie, her sister-in-law, Ms Hudson, who was living at Newbliss Farm on 13 November, and Mr Brian McGurk, a director of the respondent company and the husband of Ms Christie. It is unnecessary to set out that evidence since no issue is taken with the justices' factual findings. The only question is whether they were entitled to conclude, upon the basis of those findings, that the stop notice had not been served on the respondent.

[17] After summarising the rival submissions, the justices concluded that service of the stop notice was not effected at either the registered or the principal office of the respondent and that:

> (c) The "stop notice" was delivered in person by Mr Worrallo (Process Server) and was not handed directly to the Company Secretary (Ms Christie) at either the registered or principal offices of Keyway (Gloucester) Limited.
>
> (d) The "stop notice" was left at both the registered and principal offices of Keyway (Gloucester) Limited, with unidentified people.
>
> (e) The envelope containing the "stop notice" when left at the principal office by Mr Worrallo (Process Server) was not addressed specifically to Ms Christie (Company Secretary).
>
> (f) The applicant did not take all such steps as were reasonably practicable to effect proper service because the appellant could have sent the "stop notice" by recorded or registered mail.

[18] The question posed by the justices for the opinion of this court is as follows:

> Whether in order to prove effective service of a stop notice under the Town and Country Planning Act on a limited company it is sufficient for the Planning Authority to prove that they have left a copy of the stop notice either at the registered office or at the principal office of the company in accordance with section 233(2) and 233(4) of the Local Government Act 1972, or alternatively whether as found by the Magistrates it is necessary to prove

> that the document has actually been delivered into the hands of a Director or Secretary of the company.

[19] On behalf of the appellants, Mr Peter Wadsley submits that the stop notice had been properly served on the respondent under both section 233 of the 1972 Act and section 329 of the Act. Dealing, first, with the 1972 Act, although "person" is not defined in subsection (1), by virtue of section 5 of, and Schedule 1 to, the Interpretation Act 1978, it includes a body corporate, such as the respondent, unless the contrary intention appears. Mr Wadsley submits that no such contrary intention appears in section 233. Subsection (4) defines "The proper address of any person... in the case of the body corporate or their secretary or clerk...", thus making it clear that a person in section 233 may be either a natural or a corporate person. Thus, substituting the word "company" for "person" in subsection (2), section 233 authorised service on the respondent company:

> either by delivering it to [the company] or by leaving it at [the company's] [registered or principal office] or by sending it by post to [the company] at that address.

[20] Whatever else may or may not have been done by Mr Worrallo, he left the stop notice at both the company's registered office and at its principal office; thus, service was effected under section 233.

[21] On behalf of the respondent, Mr Dove submitted that a restricted approach should be adopted when interpreting section 233. A breach of a stop notice is a criminal offence: it is therefore vital that a corporate body should have its attention specifically drawn to the existence of a stop notice. This need is reinforced by the provisions of subsection 187(3), under which a defendant has to prove not merely that he did not know, and could not reasonably have been expected to have known, of the existence of a stop notice, but also that it was not served upon him.

[22] If a stop notice could be validly served on a company by simply leaving it at the company's registered or principal office, where it might easily be overlooked and not come to the attention of anyone in a position of authority in the company, the defence afforded by subsection (3) would be effectively nullified. He submitted that "person" in subsection (2) of section 233 meant a natural, not a corporate, person. Service of companies was dealt with separately under subsection (3). The word "may" in that subsection should, effectively, be read as "must". Thus, he submitted, the stop notice had to be given to, or served on, the company secretary in person, unless it were sent to the company through the post.

[23] I accept Mr Wadsley's submissions. The word "person" in subsections (1) and (2) of section 233 includes both natural and corporate persons. No contrary intention appears in the section; rather, subsection (4) makes it plain that "person" includes a body corporate as well as its secretary or clerk. A company may be served by posting the document to its registered or principal office, by leaving it at the company's registered or principal office, or by delivering it to the company. Since there may be some doubt as to how one might be able to deliver a document (otherwise

than by leaving it with or posting it) to a corporate body, subsection (3) provides that the document may (not must) be given to, or served on, the company's secretary or clerk. Thus, the document may be given to the secretary or clerk at an address other than the company's registered or principal office.

[24] Even if I had accepted Mr Dove's submissions that "person" in subsection (2) does not include a company and that subsection (3) is, in effect, a mandatory provision dealing with service of documents upon a corporate entity, Mr Worrallo would still have effected service on the company by leaving the stop notice at the company's registered and principal offices, since, in addition to giving the document to the company secretary, subsection (3) enables the document to be "served" on the company secretary.

[25] How is a document to be served on the secretary of a company? Mr Dove sought to persuade me that it could not be served in the ways described in subsection (2), despite the fact that the company secretary is, plainly, a natural person. I have no hesitation in rejecting that submission. The answer to the question – how may a document be served on a company secretary? – is to be found in subsection (2). The methods include leaving it at the company secretary's proper address, which is defined by subsection (4)(*a*) as the company's "registered or principal office".

[26] I appreciate that service of a document by a local authority may well lead to criminal liability, but that is no reason to adopt a strained and unnatural interpretation of section 233. Since it enables documents to be served on corporate bodies by sending them through the post, it is to be expected that companies will make the necessary administrative arrangements to ensure that the right persons within the company hierarchy see important documents. I can see no reason why the same approach should not be adopted in relation to documents that are left at the company's registered or principal office. Any company that fails to make such arrangements does so at its peril because section 725(1) of the Companies Act 1985 (which was not referred to before the magistrates) provides:

> A document may be served on a company by leaving it at, or sending it by post to, the company's registered office.

[27] Thus, any company worth its salt will make arrangements to ensure that documents left at its registered office will be dealt with administratively in such a way as to ensure that they reach the correct recipient within the company. For these reasons, I conclude that the stop notice had been served on the company under section 233 of the 1972 Act, and the answer to the first part of the question posed by the justices is "Yes".

[28] In answer to the justices' second question, it is not necessary, for the purposes of effecting service under section 233, to prove that the document has actually been delivered into the hands of a director or secretary of the company. It follows that it is strictly unnecessary to consider the alternative method of service afforded by section 329 of the

Act. Nevertheless, since submissions were made, it may be helpful if I summarise my views on this issue as follows.

[29] The structure of section 329 differs from that of section 233. There is no equivalent of subsection (1) or subsection (2) in section 233, both of which appear to deal with persons both natural and corporate. Subsection (1) of section 329 deals separately, in para (d), with corporate persons, and the references to "last known place of abode" in paras (b) and (c) are indications that those paragraphs are intended to deal with natural, not corporate, persons. The same applies to para (a), since it is necessary for there to be an explanation (such as that provided in para (d)) as to how a local planning authority are to "deliver" a document to a corporate body.

[30] I accept Mr Dove's submission that it is necessary to give effect to all the words in para (d) so that to effect service on a company under section 329 it is not sufficient if reliance is not placed upon the postal service merely to deliver the document at the company's registered or principal office. The document has to be delivered to a particular person at that office, namely the secretary or clerk of the company. Since such an obligation might well be burdensome because, for example, the company secretary might be away from the registered office, or, indeed, be deliberately evasive, the local planning authority are given the alternative of sending the document by post. I have dealt with this issue shortly because the appellants did not have to rely upon section 329 for the reasons set out above.

[31] For completeness, it should be noted that the case stated by the justices is concerned solely with the question of whether the stop notice was served upon the company. The justices were not asked to, and did not, address the second limb of the defence in section 187(3)(b), namely that the company did not know, and could not reasonably have been expected to know, of the existence of the stop notice.

[32] Neither Mr Wadsley nor Mr Dove appeared before the justices. The case deals with the preliminary issue as it was presented to the justices. A person may contravene a stop notice even though he has not been served: see section 187(1). However, it is clear that, for section 187(1) to bite, there must have been effective service on someone: see section 184(3)(b) and (6). A stop notice cannot be contravened until a date specified in the notice. That date must not be earlier than three days after the date upon which the notice was served (absent special circumstances), and must not be later than 28 days from the date upon which the notice was first served on any person. Under subsection (6), a site notice may be displayed only "where a stop notice has been served in respect of any land".

[33] Thus, for a contravention of a stop notice to be based upon the fact that a site notice has been displayed, there has to have been prior service of the stop notice on someone. The statement of case contains no information as to whether the present stop notice had been served on any other person or whether there was a site notice, and there is no finding as to whether the company, even if it had not been served with the stop

A notice, nevertheless knew, or could reasonably have been expected to have known, of its existence.

[34] Had the first part of the justices' question been answered in the negative and the second part in the affirmative, it would have been appropriate for them to have investigated these other issues. In the event, they do not arise since, for the reasons set out above, the company had been served with the stop notice.

*Stop notice served.*

www.ingramcontent.com/pod-product-compliance
Lightning Source LLC
LaVergne TN
LVHW021726021026
849161LV00004BA/736
*9780728204386*